MUSTANG

M U S T A N G

Text by

S HARON C URTIN

Photography by

Y VA M OMATIUK & J OHN E ASTCOTT

Designed by

G REG S IMPSON & E RIC B AKER

A MERICAN W ILDLIFE IN A MERICAN S PACES
R UFUS P UBLICATIONS , INC . • B EARSVILLE , N EW Y ORK

MUSTANG

Rufus Publications, Inc.
Text ©1996 Sharon Curtin
Photographs ©1996 Yva Momatiuk & John Eastcott
Compilation ©1996 Rufus Publications, Inc.

Library of Congress Catalog Card Data:

MUSTANG/Photography by Yva Momatiuk & John Eastcott/
Text by Sharon Curtin

p. cm.

ISBN: 0-9649915-1-9

1. History and Evolution of the American Mustang.
2. Non-fiction. 3. Horses. 4. Photography.
I. Nature and Natural History.

96-83438 PCN

Project Director: Richard Ballantine
Designers: Greg Simpson & Eric Baker, Eric Baker Design Associates
Production Manager: Amie Cooper, The Actualizers
Maps and Charts: Gilles Malkine

Separations by Piquet Graphics
Printed in Canada by the Friesens Corporation

Published simultaneously in the United States and Canada

ACKNOWLEDGMENTS

To PHOTOGRAPH MUSTANGS WE HAD TO FIND THEM FIRST. Our foremost thanks go to those who generously gave of their time to help us look for wild horses in the mountains and deserts of the West. We are particularly grateful to Thor Stephenson of the Bureau of Land Management, who traveled with us in winter and in summer, sharing his knowledge and love of wild horses in the sun-baked, frost-ravaged range where they seek sustenance and safety. We also thank Reverend Floyd Schwieger and Jerry Tippetts, who introduced us to the mustangs roaming the Pryor Mountain Wild Horse Range.

We are grateful to BLM wranglers Vic McDarment, Bobby Anderson, Todd Nunn, and Jim Williams, as well as pilot Ron Shane and BLM wild horse specialist Roy Packer, whose work was made more difficult by our presence. Our thanks go to Dayton Hyde, who let us explore the Black Hills Wild Horse Sanctuary in South Dakota. We offer our special gratitude to the keepers of the Spanish Mustangs, Emmett, Gioja, Josie, Aileen, and Bige Brislawn, for welcoming us into their lives.

Larry and Helen Higby kept their home open for us and helped in many wonderful ways, and so did good friends John Osgood, Maureen and Ray Sternberg, and Lauren and Zbyszek Bzdak. We are indebted to Gary Starbuck, Billy Eppler and the Horse Hill trainers of the Wyoming Honor Farm for their assistance and patience. Nick Argyros supplied us with maps and music, Kel Van Demark kept our ancient 4WD going well beyond our boldest expectations, and James Manual, Debbie Un and Errol Selkirk stocked our home fires during the many months we spent among wild horses. We warmly thank our agent Joe Spieler for believing in our work, and friends Gayle Jamison and David Hall for sharing our love of photography. Our daughter Tara, who explored many wild places with us and cares for horses in a passionate and wise way, enriched us with her enthusiasm and advice. We also know that without the inspiration and creative power of our publishers, Richard and Betty Ballantine, our long quest would not have resulted in this book.

—YVA MOMATIUK & JOHN EASTCOTT

For my mother, Freda Lee Wulff

—SHARON CURTIN

For our daughter Tara, who turned 16 among wild horses

—YVA MOMATIUK & JOHN EASTCOTT

CONTENTS

OUR HORSES, OUR LAND

L AND AND SKY ARE SEAMLESS, a dark mass marked by bits of cold reflected light, a black, empty space with ice-light above and below. This winter night there are few ways for the senses to orient the body. Hard ground beneath my feet, bright stars above. There is only a vast absence of light and shape and a wind that blows so cold and hard it steals all sound, leaving a hollowness in the air. Even my nose is disoriented; the cold and wind rob the air of smell. I am blind, deaf, bereft of all the ways I hold myself in place and time.

For miles in every direction, there is no light in a window, no traffic humming, no glow of industry. Out here, civilization does not exist. This is the Continental Divide Basin in south-western Wyoming. I stand on a great plateau, over a mile above sea level, surrounded by the invisible curve of the Continental Divide. In my mind, the Divide is like a great corral, where the ground slopes east, then west; where the water runs toward the Atlantic Ocean under one foot and toward the Pacific under the other. The wind is the controlling presence here; the wind and the absence of water in an immense open, rolling space. The Divide is an irregular mark on a map, almost insanely wandering across mountains and plains and buttes and mesas. Maps seem like unimportant human constructs out here, like civilization or progress or art. The only important question is: are you tough enough to survive?

The first people who wandered here had visions telling them life began in these rocks and mountains and winds. They came to worship and hunt, and then left. This land supported no human settlements for thousands of years. It was a place to travel through, to visit, to explore and experience. Prehistoric man left a few cryptic carvings and campsites; the later Native American tribes had a larger presence, but marked the land lightly. Early European travelers marveled at the scale of the land, the mountains and grasslands, yet for hundreds of years did what humans had always done—smoothed a few trails, killed a lot of animals, and left the small signs on land and rock of people just passing through. Life here was life on the move, everything adjusting to the paths of wind, water, rocks, grass.

Standing here, freezing in the wind, somewhere in the Red Desert of Wyoming, I am surrounded by a total and uneasy absence of any human force or presence. I begin to feel dinosaurs moving in the dark, glaciers grinding toward me, ancient chasms opening under my feet. The wind tries to strip me naked, leaving me shivering and exposed in all my puny humanity.

The wind changes and small sounds begin to register. Ice cracks under a shifting rock, bunchgrass brushes against dry scrub, dirt skitters busily across frozen ground. Now I can smell sage, juniper, dry grass, dirt. Slowly, my brain begins to count the things I know. The air is almost too full now—the wind is no longer erasing sensation, but carrying sound and smell. I can hear coyotes yipping and grousing, sharp, small wake-up sounds miles away in the rimrock. Hidden antelope cough and whoof, the soft, uneasy communication of runners, stretching and getting ready to move.

Dawn light begins to pick out small details. Ice in small depressions flashes back at the sun. Snow, raked in hard ridges by the wind, reflects the opal of early light. More colors appear; the dark green, tortured shape of a juniper, a burst of red on rimrock, shades of gray and white and brown and rust in the dirt and snow. Everywhere is the subtle, feathery gray of sagebrush. I can see nearly a hundred miles in every direction. It is a cruel, bareboned landscape, defined by distance and shaped by the wind.

As if they were conjured up by the magic of the wind and sunrise, I hear the soft cadence of unshod hooves hitting hard earth. No other creature makes that strong, firm beat on dirt; I know these are horses. MUSTANGS! First I see strong heads with sharp, pointed ears, dark forms moving out of an unseen draw, a deep cut in the ground. They come dressed in the colors of the earth around them, the mottled gray of rock, deep brown of dirt, the dusky red of distant buttes. Some are marked by white, like paint thrown on legs and face and body. These are not the celestial horses of myth, but something more elemental and primitive: four-legged, solid and strong, these creatures belong to the earth.

The band of mustangs moves from a secret shelter hidden in the folds of the plain, looking, in that instant of emergence, as if they were molded from the earth by the wind. They are led by a dark bay mare whose winter coat is thick and heavy with frost. As individuals scramble for footing, pulling themselves onto even ground, they group around the bay mare, shivering and shaking, stamping and snorting, waking up and trying to stay warm. The wind is quiet now, and the mustangs seem unaware that they are being spied upon.

The weak winter sun backlights the group, sending shadow horses in my direction. All of us—observer and observed—are too frozen to be fully alert. The mustangs look like fur-coated commuters at a bus stop. Even the very young stand quietly; to conserve energy, no movement is wasted. Winter survival depends on a kind of moving trance, a mobile hibernation. They still must graze for sparse winter feed, eat snow for water, and find shelter out of the wind. Heads down, they huddle together for warmth and comfort. In this light, the group looks like one multi-legged creature; shifting and changing shape against the snow, a cloud of warm breath marking its place.

The mares sense my presence first. Heads snap up, ears point like compass needles in my direction. The group spreads out, mares nickering softly, urgently calling their young, who move closer and shelter against the broad sides of the mares. Now I can see the band is larger than I expected; more than twenty mares and young of varying color, size, shape and sex. The dominant bay mare has a formidable presence; even from a distance of fifty yards, her authority and alert, intelligent head command attention. She stands a little in front of the group. When she moves, they shift. When she begins to snort and nicker, blowing clouds of hot breath into the air, they answer. They know I am here and feel threatened. In an instant, the band of mustangs moves from relaxed and half asleep to fully alert, grouped protectively together, communicating with each other, ready to flee. These horses, although unbranded and wild, are accustomed to sporadic human contact. The Bureau of Land Management conducts

periodic round-ups in the area; crews from oil and gas companies, and other lease holders, move over the Red Desert regularly. Primitive roads, fence lines, abandoned shacks, gunfire and blowing garbage are part of the experience of protected mustangs in open spaces. Still, they are clearly uneasy—encounters with my species usually end badly despite their special status.

They wait and watch. They are unsure how much of a threat I pose; now I am only one biped huddled against a rock. But the long history of man and horse replays. We have tracked and hunted this species, waited in ambush to catch them unaware. Their experience tells them I may be a threat, may run them down, take away their foals, kill and eat them. Today, right now, I am the only predator they fear.

The band continues to move restlessly, and now I see two adult stallions have moved forward, flanking the group like bodyguards. Either this group is composed of two bands joined together for the winter, or the dominant stallion has taken a partner. This is not uncommon. The mustangs keep shifting; they look so much like a starting line-up, waiting for a leader to call the play, that I wish for jerseys or numbers so I could identify the players. They know the signals and who is in charge. I can only guess.

The two stallions, one almost red in the sun, with three flashy white stockings; the other dark brown, almost black, a white blaze marking his face, are pacing now, shaking their heads, stopping to paw the frozen ground. One side-steps in my direction, sniffing the air, tossing his mane. I think he is trying to scare me, letting me see his size and power. He weighs over a thousand pounds to my one hundred and thirty. He is armed with four rock-hard hooves, speed, and the righteousness of one who knows he belongs; I have a gun, but in size, fighting ability, and speed, I am no match for a mustang stud. He belongs here, I don't. The two of us, woman and beast, gaze across a distance of a few yards and thousands of years of history.

The first man and the first horse must have faced each other just so; frightened, curious, destined to forever remain strangers. In that moment, a lifetime of loving horses means nothing. I discover a primitive fear of an animal bigger and stronger and wilder and more purposeful than myself. Out here, right now, the mustang rules.

The big dark stud wheels away, head extended, breaking the silence with a loud, prolonged whinny. He and his flashy sidekick gather the family, and in seconds have them running at full gallop, frozen dirt and snow flying from a hundred hooves. For a moment it sounds like a stampede, horses snorting and blowing and whinnying; the ground shakes and rings from the pounding.

Then they are gone, leaving only the echoes of their passage, their rich earthy smell, and an empty spot in the landscape.

Now the country seems smaller and meaner. I stand alone, the largest mammal here, yet somehow sadly diminished by the disappearance of the mustangs. A wise man once said this landscape needs horses. The land and the mustang make each other beautiful.

This is nature's art, to create something so perfect and complete. These are the things that make us feel more alive, richer. If the mustangs were gone forever, it would be as if the world were robbed of music or poetry or love; we could survive without these things, but a great sadness and grief would fill those empty spaces.

Surely the earth and this landscape and the mustangs and man can all be saved. We are more beautiful together; wind, sky, earth, horse, man.

Genesis

II

GENESIS

MUSTANG

A HUNDRED MILLION YEARS AGO, in Cretaceous time, the great plains of North America were covered by a broad sea from the Gulf of Mexico to the Arctic Ocean. The water spread west and east, then retreated, rising and falling, creating coastal plains and covering them. Nothing was solid or permanent; instead, it was a great swamp, flooded and revealed, a place like the Florida Everglades. Millions of years of erupting volcanoes, advancing and receding glaciers and oceans, gale force winds, earthquakes and massive rock slides followed. Yet the western landscape today—rolling flats and gullies, mesas and canyons, mountains and rivers—still shows the shape of that inland sea.

Hold a fistful of Wyoming dirt, and you have in your hand the debris of a planet creating itself—ash from volcanic activity, black shale from buried sea algae, limestone from sea shells, petrified and pulverized forests, sand—rose-colored and buff and cream—carried thousands of miles over millions of years by hurricane winds. The badlands and ridge lines look like fossilized waves, surf made solid, breaking sharp and ragged against the sky. Mountains were pulled from the center of the earth; glaciers deposited rocks and bones from the Arctic north. It is not a peaceful place. The memory of violence from its formation still howls in the constant wind.

Fifty million years ago, the seas were gone and rivers ran off the mountains into vast subtropical plains, flat and green and humid. Animal life scrambled in the lush undergrowth. There, browsing on leaves and fruit, was a small mammal the size of a dog or hare, with an arched back, long tail, short neck and legs. Only 10 to 20 inches tall, small-brained and not very agile, the animal lived in a world with an abundance of forage and an abundance of danger. This was the first equine, called *"Eohippus"*, the Dawn Horse.

We know the history of the horse from the amazing fossil record left behind, like a scrapbook kept for millions of years by a doting Mother Nature. From *Eohippus* (now called by the less poetic name *Hyracotherium*) to the modern *Equus caballus*, the fossil record traces an unbroken line, the unusually complete history of a species. In the late 19th century, there were famous exhibits of this fossil record, designed to show the scientific irrefutability of the evolutionary theory. The family tree of the horse was included in textbooks, and provided evidence for the battle between creationists and evolutionists. For almost a century, since the discovery of the Dawn Horse fossil, the horse carried the burden of scientific proof for scholars.

The evolution of the horse occurred over millions of years, not in clear, even jumps, but in spurts and lunges, in response to changes in the ecology of the North American continent. The horse we see today evolved as the great, sweeping grasslands developed; the grasslands evolved in response to the rise of the massive mountain ranges that stole moisture from the air and sent drying winds over the plains, forcing plants to develop deep root systems and a lower profile in order to survive. Sometimes there were long periods of stability, sometimes enormous, almost sudden, changes. The movement of rock and plant, animal and water, wind and sun, forms a panoramic and connected narrative, everything interdependent and changing. Because of the mountains, the climate would change, grassland prairies evolve and *Eohippus* become a quite different and magnificent creature. The landscape and the horse evolved in concert, grassland and nomadic grazing animal moving in complex and delicate harmony.

By four million years ago, true *Equus* began to evolve. The vast tropical forests were completely gone, the climate dry and windy. Huge mountains of ice moved slowly south, then

retreated to the far north, leaving clear running rivers and boulder-strewn valleys. The inland sea was eventually transformed into an ocean of grass, cut by canyons, and dotted with the occasional hardy tree. In this new landscape the browser *Eohippus* became larger and longer-legged; developing speed and agility to increase its chances of survival. Its head elongated and strong, high crowned teeth developed to grind the tough grasses. It took almost sixty million years of different stages of development for this species to approximate the form we see today. Built to run, with a streamlined body and long legs balanced on a single hard toe, they moved with an efficient springy stride over the emerging plains. Speed meant survival, not only to escape predators but also the periodic fires that blackened and cleansed the plains; only the antelope was faster than the horse.

The head was very different now; a larger brain, long nose and deep muzzle and those high crowned teeth. The horse developed stereoscopic vision, the ability to focus both on the grass in front and the danger in the distance. The sharp, dagger-like ears rotated, listening for messages in the wind. At first, they probably had stiff upright manes—the flowing mane and tail are a later development—but if you time-traveled back four million years you would immediately recognize the horse. This was a very successful species; beginning about two million years ago, and continuing through the ice age, ancestors of the horse moved across the existing land bridges and spread from North America throughout the world. Soon these hardy and adaptable New World animals were thriving in Asia, Europe, Africa, and the Middle East. Grasslands, prairie, plains, savannahs, steppes, pampas, desert; one quarter of the earth's surface is grassland, and these open spaces were marked by the sound of running hooves, the rhythm of *Equus* on the move.

Evolution never occurs in a straight line; the horse did not evolve from the rabbity *Eohippus* to the refinement of a racehorse in simple equal steps. The family tree has many branches; some successful, some not. The zebra, ass, and donkey are examples of other branches that survived. Evolution is a movement; the forms of life on earth, and the planet itself, continue to change. What endures and adapts is successful. Everything depends on everything else, moving and changing in the ecosystem we call Earth. The basic element of life, carbon, can be found in coal or a diamond or the noxious fumes from a car or in every cell of a living body; this ability to change is a sign of success, in elements, in plants, and in animals. Life, at least on this planet, is about change and mutation, a never-ending journey of survival.

In Africa, one can touch the preserved prehistoric footprints of two successful grassland species, laid down early in their development. One set is the fossil hoof prints of an adult ancestor of the horse and her foal; they show a gait called a running walk. Another set of prints planted in mud and frozen in time was placed by one of our early ancestors, walking upright and gazing across the grassland. Both man and horse evolved with the grass, mankind in Africa and the horse in North America. Both were nomadic, moving across continents and founding new colonies all over the world. Earth, grass, horse, man: they moved together under the deep blue space. We don't know much about early encounters between man and horse. Prehistoric cave drawings show a beautiful creature, drawn with great detail, moving with other running animals. In one drawing, the horse has a finely shaped head and zebra markings on its legs. It could be a picture of a modern mustang, running in the Montana mountains. We do know that early man hunted and consumed large mammals, including the horse.

With spears and other tools developed on the African plains, early man moved north, traveling with the grass to the Middle East and Asia. Like the horse, our physical refinements as a species included changes in the way we moved, an increase in size, and the development of more efficient teeth. But it was our technology, the ability to invent tools, that enabled us to survive and prosper. Eventually *Homo Sapiens* would thrive in forests as well as grasslands, becoming a species which would domesticate dogs, pigs, grains, roots; which became capable of living in the freezing tundra and the searing desert.

When this very efficient primate moved across the Bering Land Bridge onto the North American continent, it continued a journey from grassland to grassland, from Africa to the Great Plains. This exactly reversed the path followed earlier by the horse; yet both species were mov-

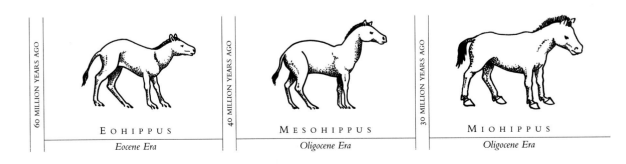

60 MILLION YEARS AGO

EOHIPPUS
Eocene Era

40 MILLION YEARS AGO

MESOHIPPUS
Oligocene Era

30 MILLION YEARS AGO

MIOHIPPUS
Oligocene Era

ing toward global occupation. About ten thousand years ago, a man carrying a spear stood in the homeland of the horse, a huge tall grass prairie stretching across to the horizon, populated by horses, camels, woolly mammoths, and antelope, the herds of grazing animals peacefully feeding. From this human's point of view, the New World was just a bountiful supply of protein. Within a tiny blink of time—probably less than a few thousand years—all the large mammals, including the horse and excluding only the speedy antelope, were extinct in North America. The camp-sites and killing grounds of early man suggest this was not caused by some cyclic change in climate; the bones of newly extinct animals are marked by spear points and human teeth.

Man came to the New World and horses disappeared. The American grasslands were empty of horses; no more stallions standing on a ridge line, no sound of hoof beats, no new-born foals in spring grass. The landscape became less beautiful.

Meanwhile, in the Old World, both *Homo Sapiens* and *Equus* continued to thrive. Man became less nomadic and began to domesticate plants and animals. The cultivation of grains—wheat and rye in Europe, rice in Asia—and the husbandry of animals, including dogs, pigs, fowl, sheep, and cattle, led to the development of controlled food sources and permanent communities. Civilizations and cultures were born. Man began to mold the landscape and plants and animals in an uneasy partnership with nature; by using a large and clever brain to create technology and shelter, man began to wield control over his environment.

Horses were the last animals domesticated by man. Until about 5,000 to 6,000 years ago, grazing herds of large mammals were considered meat, targets for the aggressive hunter. Reindeer were probably the first large mammals domesticated by man, as a pack animal and a source of milk and meat. The rudimentary technology of the harness and halter soon followed;

20 MILLION YEARS AGO

MERYCHIPPUS
Oligocene Era

10 MILLION YEARS AGO

PLIOHIPPUS
Pliocene Era

4 MILLION YEARS AGO

EQUUS
Modern Era

we had discovered a way to make other beasts bear our burdens. Remaining within the same migratory territory, reindeer were easier to tame and control than the nomadic horses.

At some point, nomadic man and nomadic horse began to discover a natural kinship. Man had the intelligence to recognize the utility of the horse; but there was something more, something that shows in those early cave drawings. There is an affinity there; both man and horse are creatures of the grass. It makes a partnership seem fated, inevitable. A man mounted on a horse looks utterly natural.

The very adaptable horse evolved into different types in the Old World. From the large Norse duns of northern Europe to the thin-skinned, high spirited, swift horses of the Arabian desert, enormous varieties of shape, size, and color developed in response to geography and climate. When man domesticated the horse, he began selecting and breeding for the qualities he valued. For the horse, natural selection was replaced by the shaping hand of man. The genetic pool of the horse, formed by millions of years of evolution, was now directed by another species. Selectively bred horses carried knights in full armor, pulled plows and carts,

ran endurance races in the desert; man is unique in nature in practicing gene manipulation on other species in order to suit his own needs.

The horseless rolling grasslands of North America saw a people who moved with the grass and the buffalo, using well-honed hunting skills and building a society based on the spear and the buffalo. The dog, a domesticated wolf, was the only beast of burden for this still nomadic people. Traveling, hunting, and butchering followed annual cycles. There were variations in social patterns among the tribes, but the basic flow of life was with the grass, the grazing herds, and the simple technology they developed to exploit it. It was a culture waiting for the return of the native nomad, the horse.

The horse did not return to North America until the 16th century. And it was a very different creature, selected and molded by man for centuries, the basic form and temperament fine-tuned and perfected. What nature had begun, and man had eradicated on the prehistoric American grasslands, came home an animal bred and domesticated by man to be a living, running, breathing instrument of civilization.

In an incredibly short three hundred years, herds of horses were again running free in their native grasslands. Beginning with the few who arrived with Spanish explorers, European colonization of the New World continued to bring horses home, and the horses responded by multiplying prodigiously. In the West, the Spanish brought breeding stock—Andalusians and other hardy Barb crosses—which proved fecund enough to eventually provide mounts for the colonizers and the Native Americans. The Horse Culture of the Great Plains was born.

The Plains Tribes took to the horse as if they had invented it, totally changing their lifestyle. Horse theft, military prowess, and more efficient hunting became the basis of the economy. The tribes that were partially settled communities left their villages and became fully nomadic, mounted on horses descended from stolen and abandoned animals from the Spanish and other European colonizers. By the late nineteenth century, over a million horses had re-established themselves in the wild. These horses came to be known as mustangs, a corruption of the Spanish *mesteno,* meaning wild or unclaimed. Over time, many different breeds of horses added to the genes of the mustangs, including the Spanish Barb and Andalusians, English Thoroughbreds and Quarter Horses, American Morgans and draft horses. Some bands of horses existed in isolation, and developed particular characteristics; others continued to breed like the men from whom they ran, taking their form from a mixed bag of genetic material.

The ebb and flow of settlement and conflict in the West first added to, then subtracted from the herds. Native Americans began their own selective breeding, choosing for color and conformation. (The most famous is the spotted Appaloosa, developed by the Nez Perce.) The U.S. Cavalry and the cattle ranchers caught mustangs, putting them to work. But even as many horses of the wild breed were captured for man's service, the mustang herds continued to grow.

And as the mustang multiplied, so did the myths of the American West. The nation spread out from coast to coast, sending explorers and trappers, soldiers and settlers, gunfighters and preachers, outlaws and families. The national narrative, the story we all share, is based on

the high plains drama; the man, the horse, the endless sky and grass. The story always begins

with these relationships—with the conflicts between the free nomadic society and the needs of

a rapidly expanding population. The romance of the wild, represented by the noble Native

American, or the rogue mustang, or the cowboy and his horse, is pitted against the rational

force of progress, in the form of agriculture, or mining, or the railroads.

As the narrative played out, the Native Americans and the buffalo and the mustang

retreated; the cowboy gave up his spurs, the wide open spaces were tamed by barbed wire.

Eventually the heroes of our story lost their land and their freedom and usually their lives to

the new civilization.

By the beginning of the 20th century, the buffalo were almost extinct, the Native American reduced to a bare subsistence living on small reservations. Cowboys and cavalry soldiers held on longer, but their utility began to fade as well. The men and animals who moved with the wind and grass lost their meaning and their way, mowed down and overwhelmed by machines and cities and conformity.

The romance of the Horse Culture lasted only a few hundred years, but the power of its narrative lingers. We haven't developed another myth, another legend, strong enough to obliterate the natural union of earth, sky, horse, and man moving together.

Less than a hundred years ago, millions of wild mustangs still ran free in the western United States. By the time the railroads came, and barbed wire made fencing large holdings economically possible, the grassland itself was under siege, attacked by the iron plow and a nation's need for cultivated grains. Yet mustangs managed to survive into the twentieth century, mainly in rough, semi-arid country the new agrarian culture rejected.

By this time, the mustang had developed certain characteristics, becoming a breed or type of horse. Never large, it was usually around 14 hands high, with a short backed, strongly muscled body; most of all, the mustang had strong bones and hooves. Colors varied, from grulla and buckskin, with the strong dorsal stripe and sometimes zebra-like markings on the legs, to showy palominos and pintos. In some locations, isolation created a recognizable type—for example, mustang herds in the Pryor Mountains on the Wyoming-Montana border and the

Kiger mustangs in Oregon. In general, mustangs were not always elegant or impressive. Their hardiness, intelligence, agility, strong bones and feet, and endurance, were qualities admired and needed by nomadic man. Once the culture changed, mustangs were as unwanted as the buffalo.

Wild mustangs were regarded as a nuisance by the emerging cattle industry; the grass was meant for cows, not wild animals. Mustangs retreated to even higher ground and more isolated country, away from the forces of civilization. Shaggy, stubborn, sometimes stunted because of poor grazing conditions; carrying the wild colors of long dead ancestors, mustangs became an object of scorn for people who wanted their horses slick and predictable.

HOME ON THE RANGE
Free-Roaming Mustang herds

Yet the mustang had one last call to mankind's service. During the Boer War, hundreds of thousands of mustangs were rounded up and sold to the British, who demanded horses with endurance, courage, and agility. And again, during World War I, more mustangs were called to serve on the battlefield. In the first years of this century, over a million mustangs went to war. None came home.

These mass round-ups created a new professional, the mustanger. For decades, using horses, trucks, airplanes, and dogs, these hunters tracked down, rounded up and shipped countless mustangs. When the military stopped needing horses, another killer market was born—mustangs became a source for the American pet food industry. Within fifty years, by the 1960's, mustangs had been reduced in number from over a million free running horses to less than 20,000.

This strong symbol of freedom, the perfect nomad, man's loyal partner, an evolutionary miracle—no matter; mustangs were dog food. Until 1971, systematic removal of the mustang, especially from public land, allowed revenue producing livestock unimpeded access. If stockmen couldn't sell mustangs, they regarded the fleet animals as just another pest, and had them killed out on the range, in midstride. Like Buffalo Bill and others who tried to exterminate the buffalo, mustangers bragged about the numbers they caught and sold to slaughter, or killed on the hoof. The mustang had no protection, no refuge, in the West.

FAMILY

Harem Bands

THE WIND CARRIES THE SMELL OF FROZEN TUNDRA, ancient glaciers, the threat of slow, inevitable death. In all the western landscapes where mustangs run, winter is more than brutal, more than harsh. From the Pryor Mountains on the Montana-Wyoming border to the canyon slashed Nevada deserts, the combination of capricious, violent weather and seasonally sparse grazing tests the bottom of every living creature. Yet the mustangs endure, standing planted like rocks through howling blizzards, eating snow for moisture, searching out bunchgrass, salt bush, even twigs and bark to give them a few calories of heat.

Five hundred years after horses returned home, first with the Spanish, later with other Europeans, they still endure, hiding in places human beings rejected, forming free-roaming herds. Qualities that contributed to toughness and endurance were tested in a natural laboratory; only those mustangs which survived reached maturity and in their turn produced progeny. In nature, the success of an animal is measured by the ability to reproduce its image, generation after generation, adapting and modifying, carrying a genetic code into the future.

The code of the mustang is the code of the nomad. Today's survivors in the American West are the hardiest descendants of a naturally tough and resilient animal. The ghosts of all those millions of horses who once roamed the grasslands of the world gave the mustangs a legacy of physical qualities—strength, speed, agility, keen senses, intelligence—and the patterns of behavior that increase their chances of survival in some of the most rugged country in the world.

Mustangs, like all nomads, exist only in the context of their environment and their relationship to each other. They are shaped, defined, nurtured by the landscape; and they naturally form families, called bands, and communities, called herds. Since mustangs are not territorial, all they defend is the integrity of their selected family. They are completely social animals, at home only in the group. They have developed and observe a complex web of relationships, a way of being and functioning that helps them survive.

The foundation of the mustang world is in the harem band, composed of a dominant stallion, mares, and juveniles. Every member of the group has a place, a role, in the hierarchy of the family. Like any family, there are variations, some seasonal, some geographical, and some governed by the personalities and abilities of individual members of the group. For example, the aggressiveness of the stallion in seeking new mares varies; the authority of dominant mares depends on factors like age and size, as well as personality. Some groups seem to display more affectionate grooming, while other groups display irritation, jealousy, and competition. Some mares form strong bonds with each other. But for all the individual differences, mustangs seem to be born knowing they belong to each other and to the land, and nothing else is important.

Outside of the band, lost or exiled, the mustang is truly homeless; it doesn't matter if there is adequate sustenance, he cannot really exist apart from his fellows. The saddest sight on the open plains is a solitary mustang; alone and separate, he wanders like a lost and doomed soul, trying to find his image in the dust. Stallions who have lost their harem bands to younger, stronger males; mares who have lost contact; juveniles who have been driven away: they all try to reconnect. Mustangs cannot bear to be without their kind. Their pathetic search can be read in the way they move, the sadness of their carriage, even the aimlessness of their tracks. Solitary mustangs do not usually live long; the soul has gone out of them. They are a herd animal; they need each other.

During the winter, bands of mustangs concentrate their energy on surviving the weather. Some years, the death rate can be enormous among lactating mares and foals; almost none will survive a truly harsh winter. Capricious nature will also send a milder season, even in the mountains, and every mustang will make it to spring, gaunt and depleted but alive. The band stays together, huddled against the wind, moving only to graze and find shelter, waiting for spring. The sexually aggressive behavior of the stallion disappears; in some bands, his leadership role is secondary to that of the dominant mare, who will lead the group out of the wind, to grazing, to find water or snow. Winter is a prolonged test of each individual's endurance. Only the strongest and luckiest will see spring.

With spring, everything changes. Mustangs begin to travel in tighter bands, as the stallion senses the beginning of breeding season. He shows more aggressive herding behavior, and no longer permits the members of his family any distance. He spends more time investigating the mares, smelling and sniffing, checking their urine and droppings for signs of their readiness to breed. The mares begin to be less tolerant of each other, and of the juveniles. In fact, where winter saw the mustangs almost comatose in their need to reserve strength, spring is marked by tension and activity.

Most of all, spring means birth, renewal, new life. Most mustang foals are born when the grass is just showing, the days just beginning to lengthen. After an eleven month gestation period, carrying a large and hyperactive fetus, the mare usually moves away from the band to

have her foal in peace and quiet. Not far away, just within some understood limit of safety and
privacy, this is the one thing a mare will do alone.

And she produces a miracle. The violence of birth, the sudden appearance of new life,
the combination of water and grass and sex, creates an entirely unique creature. The first hours
after birth, mare and foal become acquainted. She cleans her offspring, imprinting and mingling
their odors, so they will forever know one another. Within hours the foal stands, struggling to
balance on disproportionally long legs, the legs of a born runner. The mare encourages the foal

with soft nickers, nuzzling and licking, nosing and pushing. Instinctively, the foal moves toward food, suckling the first precious milk, seeking warmth and protection and nourishment. During this time, she and her foal will form a strong and unbreakable bond and learn to recognize each other by sight, sound and smell; she assumes her role as mother, the foal becomes a mustang. He is born to move and to live in the company of other mustangs.

Within a few days, mare and foal rejoin the larger family, who have remained nearby. The mare is very protective and possessive at first, but soon allows other members of the band to approach and inspect the youngest member. Although the bond between mother and foal remains the strongest—her call will always evoke a response from the foal; his sounds of distress or hunger will always stop her in her tracks—the foal becomes part of the community.

Years ago, in the spring, I remember watching a band of mustangs which had been devastated by a harsh winter, a round-up that had taken some of their youngest and healthiest, and further aggravated by an increase in human interference because of mineral development in the area. The mustangs seemed rudderless and confused, although a young and apparently vigorous stallion was present. Even though the weather was warm during the day, they were inactive, standing huddled together, meekly accepting sun, rain, snow, wind, the last to graze new grass or take a place at the water hole. They were so dispirited I thought they were ill, infested with parasites or poisoned or toothless.

Looking at this bunch, I thought I finally understood why some people felt they would be better off dead, that being dog food was preferable to such miserable existence. And yet, there was plenty of new spring forage; they should be recovering from the stress of winter by now. They were wildness diminished, hunkered down, barely breathing. They even ignored my presence, displaying none of the lively nervousness of the truly hunted. They looked as though they just didn't care.

As I watched one mare raised her head. Then another. The stallion moved forward; and suddenly all of them turned toward a single horse moving out of the sage. The mares began to nicker, then whinny, milling around as if forming a welcoming committee. They didn't move toward the solitary animal; they waited, blowing and snorting, somehow brought awake by the new arrival.

She was a little sorrel mare, ragged in her winter coat, ribs and hip bones moving under her hide, too thin, I thought. But trotting at her side was a new foal, an awkward colt mustang, as shiny and smooth as a copper pot. Unsteady and uncertain, but full of curiosity and trust, he moved with his mother toward the remnants of his family.

Meanwhile, the band was transformed. Hesitantly at first, one of the older mares moved forward, making soothing sounds, asking permission from the mother. She was followed by other members of the band, including the stallion, and they all stood marveling at this new life in their midst. The colt accepted this adulation as his due, alternately hiding behind his mother and boldly rushing forward to smell these new acquaintances.

During the next few hours, I watched the little colt save his family. His appearance seemed to give them pride and hope, and enlivened the entire band. His mother objected when

a larger mare got too close; the dominant mare snapped back, an energetic action unlikely an hour before. Everyone began eating; the yearlings formed a separate, playful little group. They began to look and act like mustangs.

But the best part happened just before dusk. The stallion began grooming the new mother, the first grooming behavior I had seen all day. Quietly, he moved close to her, and the sorrel leaned into him, and they began to rub each other's necks, scratching backs, pulling out old tufts of winter hair. At her feet, the colt rested, nose down in new grass, exhausted by the family reunion.

When the grooming ritual was completed, the mare moved away and emptied her bladder, apparently relaxed and happy. And the stud walked over, smelled her urine, straddled it, and covered it with a strong stream of his own.

Life goes on.

Stallions

S EX AND FAMILY. These define a mustang stallion's universe. Unlike other large mammals, he is not interested in protecting or acquiring territory and he is not indifferent to the family he produces. Male deer, for example, show an interest in female deer only during the rutting season; they assume no role in protecting or rearing their progeny. They will move and gather in herds to migrate, but the family unit is the female and her young. The mustang stallion takes his responsibilities seriously; polygamous, he keeps his harem band together, serving as mate, protector, and sharing in the care and education of foals. The stallion as a family member is altogether admirable—in his single minded devotion, he cannot be distracted. His focus, his life, is his family.

Loyalty, vigilance, courage, persistence: these aspects of a stallion's behavior seem virtuous. A biologist would say that the stallion's behavior is just the result of an instinct to preserve his gene pool, that by protecting the mares and their young, he guarantees the continuation of his DNA. He can retain his sexual partners only by fighting off predators and other stallions; by staying with the family he makes sure nothing happens to interfere with his breeding rights. He is driven by his instincts to be fiercely possessive and domineering; on guard through the seasons, against all enemies, the stallion invests his life in protecting the integrity of his harem band.

Late one November afternoon, I am sitting on the tailgate of a truck, tired out from bouncing around, watching a small band of mustangs. In the clear light, they are almost sorry-looking, their colors muted and muscles hidden by thick winter coats, once graceful legs stubby and hairy, manes and tails matted with burrs. Some have such thick coats that you cannot even guess at their conformation; all the bone and line is hidden, any beauty surrendered to winter's imperatives. Even their muzzles are hairy; what could be fine, small heads look like something more primitive, out of another time. The weather in this country is brutal and apt to change without warning; in the last few weeks, the temperature has ranged from 20 below zero to today's sunny 55 degrees. All of us are enjoying the gift of these last days of Indian summer but the mustangs are dressed for winter. A sensible and adaptive creature, mustangs conserve energy in the winter, using stored body fat and available forage to maintain body temperature. Although it is late November, this group is still in good condition from a summer's grazing. The harshest weather lies ahead.

They are relaxed enough to allow me to approach fairly close, within a hundred feet or so, as I slowly move up and sink to the ground. I've noticed that the pick-up truck makes them more suspicious; maybe it's the gun rack. I've left the truck half hidden behind some scrub and sit in the sun watching the stallion watching me.

He is always on guard. No animal is as unremittingly alert as a mustang stallion. While grazing, he never takes more than a few bites without raising his head, looking around, sniffing the air. He moves from high ground above the family to a roving position on one side or another. His eyes are focused on his charges, his surroundings, me. The mares seem to ignore me; a few juveniles are overcome with curiosity and come within a few yards, dancing back and forth, sniffing and blowing. The stallion, on the other hand, displays neither acceptance nor interest. He is simply totally aware.

Especially at this time of year, stallions do not move their bands unless the threat is over-whelming. The conservation of energy is a first priority in the cold weather. As long as I remain quiet, he doesn't seem to find me dangerous. His state of alertness is constant; whether I am there or not, he can never let down his guard. In the end, I think I am more uneasy than he is; our roles of observer and observed seem to switch. I am more accustomed to watching than being watched.

The mustang stallion is a big, almost black horse. In warmer months, I imagine his coat glimmers pure black; now, with his thick winter growth, he looks almost rusty. He is young; I can see the solid muscles and strong neck, no ribs or hip bones protruding. He has few marks or scars, just a jagged line running across his muzzle, a larger scar on his shoulder. I think he may be one of the lucky ones, a young stallion who gained his harem by stumbling on a group whose leader had died. None of the juveniles resemble him; this could be his first year as a dominant stallion.

Mustang society is a sort of rotating dictatorship, with the dominant stallion assuming leadership and sexual exclusivity. When this stallion acquired his harem, he began a pattern of expelling any males who reach sexual maturity. This is a brutal experience for juvenile males—the most social of animals, banished from home, ignored by his mother and cohorts, doomed to loneliness. The outcasts try, over and over, to rejoin the band, and are chased, threatened, bitten, and driven off with increasing force. The harem band seems deaf and blind to the language of the outcast; it is as if everything that made him a member of the group has devel-oped a new definition, a new language, and his sexual maturing has made him a foreigner. It is a sudden and almost cruel transition.

Eventually (and hopefully) the outcast will join other immature stallions, forming a bachelor band. They come together for company, for protection, even for play, but this asso-ciation is neither permanent nor satisfactory. Sometimes older stallions who have lost their status as leaders will join a bachelor band; some of the bachelors will never be aggressive enough to build a harem. There is a sort of brotherhood of stallions in some places, where a few old and incapable male mustangs keep each other company. In the wild, the most impor-tant thing about the mustang is this: mares and their young provide the goal and the glue for the mustang.

I imagine my stallion was chased from home when he was two or three years old. He is big, standing about fifteen hands. His head is long, somewhat Roman-nosed, with a white blaze face that makes him look startled, even a little goofy. His body is strong, with a nice deep chest. I think he must be about seven or eight years old. I notice, not for the first time, that I tend to regard any mustang I am observing as "mine." This may be a way of 'identifying with' the subject, a way of feeling intimate. But part of it is that I grew up out in this country and can still remember when any free running horse, unbranded, was considered the property of

whoever could track it, corral it, rope it, and get it home. Now mustangs are public property, protected by Federal law. This black stallion belongs to the American people. But he knows he belongs to the harem band he protects.

Above and around us the mountains are covered with snow, a white mantle that will last from early fall until late spring. For the next few months, mustangs will winter in valleys, seeking shelter against the rocks and amongst the few trees, eating whatever they can find. Year round, they graze 12 to 16 hours a day, always moving, always eating. During really bad weather, they simply stand and endure.

The stallion and I continue our mutual scrutiny. At one point, he raises his head as high as he can, drawing back his lip a little and glaring down his nose in my direction. He lowers his head, shaking it and snorting, blowing in disgust it seems, pawing the ground. Lifting his feet high, bobbing his head like a show horse, he trots in my direction, wheels, and moves away again. This is challenging behavior; it is as if he is asking me: why are you here? He doesn't seem to have any personal interest; he just needs to know if I am a threat to his family. Sometimes he moves over to one of his band, rubbing necks, touching noses, nuzzling a dark yearling dozing in the grass. He looks as though he is counting them, asking if everything is okay, do you want anything? I know this is ridiculous, but the intensity of his constant attention and awareness makes him seem concerned for their well being. This winter, if it continues a normal pattern of freezing cold and high winds, some of the lactating mares and many of the foals will die. The mortality rate depends on the weather. Nature will decide; no one comes to feed or shelter these mustangs.

In the spring, the stallion's vigilance over the survivors will intensify. When breeding season begins, the scent of a mare in heat will attract competition. Bachelors want to begin a

harem and few dominant stallions have more mares than they want. A stallion will become more aggressive, keeping the band closely bunched, forcing stragglers to keep up, using his powerful body and strong teeth to bully the family. Every stallion with a harem seems to know he risks unspeakable loss. During breeding season, stallions barely eat or sleep; they are completely focused on the moment a mare signals readiness.

The stallion will begin to cover the urine and droppings of the mares with his own, trying to cover the invitation of rising hormones, but the spring breeze will carry the message. The competition begins. Every bachelor stallion becomes a stalker, haunting established harems, looking for strays, desperately following the scent of the mare. Bachelor bands continue to travel together but are now acting as individuals. They are no longer members of a team; they are sexual athletes, ready to collect trophies.

Stallions engage in battle for the sole purpose of protecting the integrity of the family. In this mustang culture, bachelors are all home wreckers. Any male mustang approaching a harem band will be challenged by the dominant stallion. He is now on 24 hour duty. He spends more time grooming and communicating with the mares, looking for all the world like an anxious suitor. He must be ready to mate when she is ready. He must keep the mares together, or risk loss. The rituals and behavior have a universal resonance; male and female exchange glances, stay close together, even show signs of jealousy; the male exhibits much macho posturing, remains constantly aware of any signals from the female, and is ready to fight any perceived interlopers.

Bachelors engage in mock fights from a very early age, building strength and skills, and are instinctively highly motivated to engage in combat. Sometimes, they will back off from a fight when faced with a determined, aggressive and clearly larger stallion. They are totally opportunistic. If a band stallion is distracted, or a mare suddenly becomes separated, a bachelor will immediately move to claim her. A young stallion will sometimes repeatedly challenge a particular leader. The bachelor can always back off. The stallion with a harem band is always prepared to fight to protect it; he may lose the fight, but he has to accept the challenge. To lose his family, his status, to be deposed, must be unthinkable, like death, for the stallion.

In the spring, the sound of a stallion's challenge can be heard miles away. Like the bugler's notes, it is both a warning and a call to battle. When competing stallions meet, they seem to grow and thicken, raising their tails and arching their necks, creating a formidable profile. The preliminary exchanges vary, but involve prancing back and forth, sniffing, sometimes defecating and smelling one another's droppings; they seem to exchange insults with whinnies, nickers, whistles, screams. Battle begins in a rush, walking on hind legs and striking with rock-sharp forefeet. Mouths open, they go for jugular veins, muzzles, tendons, ears; screaming, they throw all their weight and strength into combat, kicking, biting, tearing. There are no rules. The fight continues until one is badly wounded or surrenders by running away. The victor might chase the vanquished, but his attention must immediately return to protecting the harem. The mares generally ignore the drama and noise and blood; they, too, have an interest in the survival of the group and are not bound to the stallion.

Mustang stallions can be thought of as ludicrous, testosterone-crazed, instinct-driven, out of control sex maniacs. Even, as my horse-smart friends say, unnecessary, ugly and worthless. And yet—sitting here, watching, imagining the life of a mustang—I can't help admiring him. He is magnificent; and I know I am responding to his blunt, basic maleness. His code, or being, or soul, or spirit, is purely masculine and evokes an unexpectedly strong, almost physical response, from me. His basic sexiness is somehow enhanced by the nakedness of his need to possess and control. It creates a tension, a state of muscular alertness and pure physicality, that commands attention. The mustang stallion has authentic power and authority.

His reign is limited, however. Watching the big stud pace the heights is like watching a star who knows his days of ruling supreme are limited. The mustang family will endure; the king will be replaced. As an individual his power and glory are fleeting. It is hard to imagine; after all, he weighs at least 900 pounds, but I feel his presence is fading as I watch. The light is fading, too, and the mustangs are bunching up, sheltering each other from the cold. The day has been seductively warm, but winter comes every night. Silhouetted against the sunset, the stallion stands quietly, mane and tail flying, dark horse against the flaming, fading sky. He is so much a part of this land. Solid, natural, beautiful.

I turn back toward my vehicle, a squat, ugly, hunk of metal waiting to take me away. A necessary means of transportation, it requires enormous and irreplaceable resources to build, to fuel, to maintain. I climb inside and am swallowed up by the machine, shielded from the wind and cold, gazing at the mountains through distorting glass.

But today I watched a mustang stallion. I experienced a small taste of nature and history and wildness. It is a bittersweet gift; I know I am a symbol of everything that goes wrong here in the West. We are a catastrophic species; we insist on our right to possess, conquer, control, destroy. Today, a mustang stud showed me how much is lost because I am unconscious of my natural self and my position as a part of nature, of the earth. The mustang stallion is just what he is. He has a position and a purpose in the endless stream of life, playing, fighting, eating, drinking, breeding, and dying on the canvas of this landscape.

Bachelor Bands

ALE FOALS ARE BORN TO PRIVI-
lege—they are allowed to nurse
longer, giving them time to grow
stronger and bigger on the rich milk of their
mother. They occupy a special place in the structure of the family, where their natural curiosity
and sense of play is tolerated and even encour-
aged by the group, especially if they are the progeny of a dominant mare. All foals are protected
and watched over by the adult mustangs, but the
most aggressive foals, the primary troublemakers
who challenge established order, are generally the
little colts, the male foals.

For the first year, the colt is treated like a
prince. Colts begin to establish their special position early. Because they are larger than their
female cohorts, they are the ones who tease the
adults and other juveniles, who sometimes push
ahead of the usual order at water holes, who even
try to steal choice forage from the stallion. They
begin to establish shifting hierarchies among the

juveniles, traveling and playing and shoving in packs. During spring and summer, when the grazing is good and water adequate, colts live in paradise.

Colts begin to lose that paradise when the dominant stallion senses their rising hormones. Behavior that was tolerated begins to draw a more hostile response; the confused colts find themselves the target of bites, kicks, and shoves, forcing them to recognize a strict code of behavior. It is pathetic to watch a young mustang try to figure out what he has done to deserve such treatment, as if he is suddenly being carded at the door and refused admission to the party.

By the time he is two-to-three years old, he will be exiled, shunned by the only family he has ever known, driven off by the harem stallion. Because mustangs are totally social animals, the young males find each other and form bachelor bands. Exiled because of their rising testosterone, they form prairie gangs—hanging out together,

practicing the skills they will need to survive and gather their own harems. They still have to graze and find water and shelter; but there is an air of expectancy about a bachelor band, a feeling of unfinished business.

Sometimes older, defeated stallions will join the bachelors, like old warriors seeking out their lost youth. A loose congregation of males, the bachelor band will recognize one young stallion as leader. He will be the first to start mock fights, the leader in early dawn runs, the first one to drink at the water hole. Eventually, he will be the first to leave and begin his own harem band.

Where there are large herds of mustangs, the bachelors are easy to spot. They crowd together, finding comfort in physical contact with each other, always alert and watchful. Some look like dudes in a bar, constantly shouldering and shoving, posing and posturing. Some just have a stillness, a quality of waiting, like Romeos looking for the perfect Juliet.

Most of all, they are looking for the paradise they knew as colts—they all want to return and be part of a harem band, but this time as dominant stallions. So they wait, athletes on the bench, poised for destiny's moment.

Grazing

I T LOOKS LIKE A GATHERING OF GRASSLAND SAINTS, a community diverse and sacred, existing in peaceful glory. A herd of elk moved through here yesterday, startling the mustangs and cattle. A small group of mule deer, does and yearlings, and the ubiquitous antelope have been present most of the day. Bright sun highlights the mustangs in the valley below, emphasizing the colors and forms they carry. The other wildlife carry quieter camouflage, blending into the shadows and brush. Adaptation to climate, geography, as well as survival skills, can be read in the way the animals move, the shape of heads, bodies, and feet.

Mustangs are constantly on the move, traveling ten to fifteen miles a day, grazing on the run. They require about twenty to thirty pounds of food a day, a movable feast of different forage species. Mustang digestive systems are not efficient; nature equipped the animal with strong teeth and the ability to move and eat at the same time safely, in order to encourage a varied diet and adequate nutrition. Since they digest forage incompletely, the seeds of the rangeland are redistributed in their droppings. This valley was probably grazed and reseeded by the ancestors of the animals I am watching. Today they share the fall grasses with migrating deer, elk, and antelope, and with domestic cattle. The natural scene is peaceful; elsewhere, the political battles over who will remain on the range and in what numbers, rages.

Over fifty percent of the land in eleven western states remains public, held in common ownership by and for the American people, and managed by federal agencies. This is over 300 million acres, twelve percent of the total land in the continental United States. This grassland heart of America is now charged with supporting what law and custom deem the principle of multiple use. The western states, the original home of the horse, is a landscape strained to its ecological limit, less peaceful than it looks because we expect it to deliver too much.

Mustangs can and must share the range with wildlife and domestic animals. Scarce forage and water are allocated by federal managers, calculated in animal units (AUM) per month.

Each grazing area is assessed and assigned a specific population of grazing animals. Big game populations are controlled through hunting, which is also a profitable business in the West. Cattle and sheep ranches holding leases are told how many animals their leased lands will support. The optimum population of mustangs is determined, and excess animals must be removed.

Each of these animals has economic utility—except the mustang, which must rely on the kindness of strangers rather than dollar value. Each competing interest constantly presses its case for a bigger share of public land. More game animals, like elk and bighorn sheep, mean more income for the local economy from tourists and hunters. The stock growers are also a powerful lobby, pushing to the head of the line, historically entitled to the use of public land. Ever since the 1800's, the destiny of the west and the cattle industry have blended. Even though less than two percent of the total domestic livestock population grazes on public land, in our imagination, the West includes cattle.

Mustangs, domestic animals, and big game animals are not the only creatures subsisting on public lands—but they are the biggest living users of scarce forage and water. As a group, they eat similar forage. Because of their presence, parts of the western rangeland are on the edge of ecological collapse.

However, the largest presence is that of cattle. Cattle on public land outnumber wildlife by more than 5 to 1, and the beef industry has forever changed the grasslands. Overgrazing has resulted in a reduction of all species, including plants, mustangs, elk, and bighorn sheep. Government agencies have supported the elimination of predators in order to protect livestock, targeting mountain lions, bobcats, coyotes, foxes, and wolves, removing virtually all predators from large tracts of public land.

Mustang populations, like other wildlife groups, have been pushed to the edge of the table, forced to exist in areas of sparse vegetation and limited water, subjected to extreme weather and a subsistence diet. Once the plains were opened to livestock, the balance shifted forever for all other species. The diet of the mustang and other grazers differs enough to allow tolerance; they are not basically competitive to one another. For example, antelope prefer forbs and shrubs to grass; their diets overlap by less than ten percent. Mule deer and elk are migratory and only seasonal companions to mustangs. But cattle and sheep compete with all wildlife for space and forage.

Once the western United States supported tens of millions of buffalo. We killed the buffalo to make room for our vision of a pastoral nation, with domestic animals under human control. The land that supported countless buffalo is no longer able to provide for all the mouths at the table.

I can hear cattle lowing, and an occasional meadow lark who hasn't yet flown south for the winter. Only the cattle whine for food, standing in the brush, waiting for some human to bring water, hay, grain. The other grazers I see are quiet, the elk gone now, the antelope hiding in the sage, and the mustangs moving across the valley, eating in the sun.

Colors

USTANGS, LIKE THE BIBLICAL Joseph, wear coats of many colors. They can be spotted or pure black; palomino gold or steel blue; dapple-gray or strawberry roan. Their faces can be marked by white stars, flashy blazes, or narrow snips of paleness. In some isolated areas, they may share one predominant color—sometimes the shades called primitive, duns of buckskin yellow, mouse brown or grulla blue, with black points (mane, forelock, tail and lower legs), a dark dorsal stripe, and the legs marked by faint zebra stripes. Other herds are completely multi-hued; bays and grays and Appaloosa spinning an earth-toned kaleidoscope.

Colors are genetic; offspring inherit from their parents the genes that determine the colors they carry for life. A dominant gene will prevail; gray, for example, is dominant to all other colors, followed by bay (brown coat with black mane, tail, and legs), brown, black, and the dark red-gold shades of chestnut. Certain breeds of horses are likely to carry genes for particular colors, bred by man for both color and conformation. The mustang can show genetic lines stretching back to the spotted horses of Cortez, the dapple-gray of English carriage horses, the elegant gleam of Thoroughbred black.

Among the wild mustangs of the American West, a certain anarchy of color prevails. The pale gold of palominos marks some bands, while black is the color carried by their nearest equine

neighbors. The Appaloosa, a spotted horse bred by the Nez Perce, is still seen on the open range. Nature doesn't seem to have a favorite color for the mustang; the genetic code governing color is unimportant in the wild. Strength, speed, and intelligence are more important.

While science explains the laws governing hereditary characteristics, in nature it seems that colors and markings are more arbitrary, almost as if the choice were governed by aesthetics or whim or a sense of history. A herd of mustangs can be a living lesson in the history of the horse, a color chart drawn in the Nevada desert or Montana mountains. The colors show the range of nature's original palette and the colors preferred by fashion and fad among the people who selectively bred their ancestors.

When you see them run, you see the color of Chief Joseph's spotted mount and the sacred black mare of the goddess Demeter; the white ghost horse of Druid worship; the champion chestnut Thoroughbreds; the buckskin Indian pony. There is the red dun of the Lascaux cave drawings, and the delicately shaded horse of ancient Chinese porcelain. Every color has a story. And they are all the colors of the earth and sun; marvelous and anarchic and surprising, the colors of a crazy quilt, without pattern or plan but perfectly beautiful.

Shape

THE PROCESS OF NATURAL SELECTION and adaptation drew the basic form of the horse. About 5,000 to 6,000 years ago, human intervention began to shape different breeds and types of horses. The contrast between the heavy draft horse (thick, broad, with large feet and feathered legs) and the Thoroughbred (narrow bodied, long legged, gracefully proportioned) shows the depth of the equine gene pool.

Mustangs bear the bastard's mark in the horse world. As a group they are not a breed, although the Brislawn family of Wyoming has been selectively breeding and preserving a strain of early Spanish horses, and founded the Spanish Mustang Registry in 1957. Other groups, anxious to conserve the mustang, form a variety of societies and define the mustang in various ways.

But the mustang remains as elusive as the wind, difficult to capture and describe. In the vast open spaces of the west, it has no particular shape or conformation except as governed by its ability to endure and survive.

Mustangs are simply the toughest and strongest descendants of various breeds of horses which returned to North America with European colonizers. The original wild herds that created the western Horse Culture were descendants of Spanish horses brought by the conquistadors. By the beginning of the 20th century, when a million wild horses roamed the west, the Spanish blood had mixed with horses brought west by the colonists on the move across the continent. Heavy draft horses bolted away from wagon trains and joined wild herds; the Army

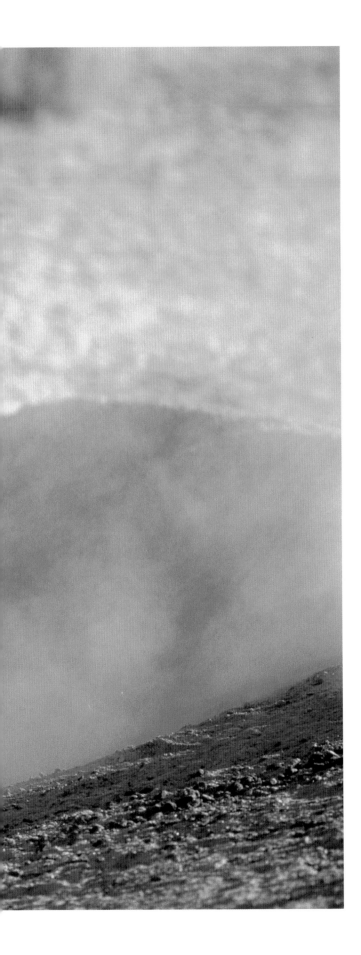

brought—and lost—cavalry horses, Thorough-breds, Quarter Horses, pacers and parade palomi-nos. Any horse that escaped, survived, and mated contributed to the mustang gene pool, the wild, tough, herds of free-roaming horses.

The idea of 'breeds' is a notion of monar-chy; bloodlines established and recorded by some authority. Mustangs don't lend themselves to concepts of royalty; they form a sort of equine democracy, a very particular American horse.

However, the shape and color and form begun in those Eurasian breeds and their influ-ence can be seen in today's surviving mustangs. The qualities are tempered and molded by the environment, and diluted by the blending of dif-ferent breeds, but you can see the shadows sug-gested in the mustang's shape.

Generally speaking, today's mustangs are a type of horse defined by geography and the 1971 Act of Congress protecting them. The approxi-mately 50,000 mustangs that remain on public land are considered to be part of the precious heritage of the American people. They are simply a living legend.

Sometimes the truest tales begin in the bones. Mustangs are generally small, because the grazing available is poor in the areas where they are safe. For the same reason, mustangs do not reach full maturity until they are four to seven years old, and their bones are stronger and denser than those of domestic horses. The hooves, the characteristic equine single toe, are correspondingly thick and strong, so durable that captured mustangs rarely require shoes. The size and shape of each individual mustang's hooves reveal ancient bloodlines—the delicate feet of the Arabian or the heavy dish-shaped feet of some long ago abandoned farm animal.

There is no typical mustang head—but many mustangs have fine, neat heads with wide set and intelligent eyes. The ears may be smallish, and the mane and tail long and impressive. The closest recognized breed in size, shape, and agility is the Quarter Horse; and many mustangs look like perfect cutting horses, strongly muscled, short backed, quick.

These are, of course, generalizations; you may as well try to describe any American crowd and face the same problem: too many variables, too complicated a mix. And sometimes, you see an animal so unhandsome, such an unhappy conglomeration, that it will surprise you and make you avert your eyes—and be ashamed.

Yet even a truly ugly, Roman-nosed, toad-eyed, thick-necked, sway-backed, mud-colored mustang can be beautiful where he lives and survives and endures in nature, running free. Lineage, conformation, color, size, don't matter in the wild. Only toughness and endurance and character count.

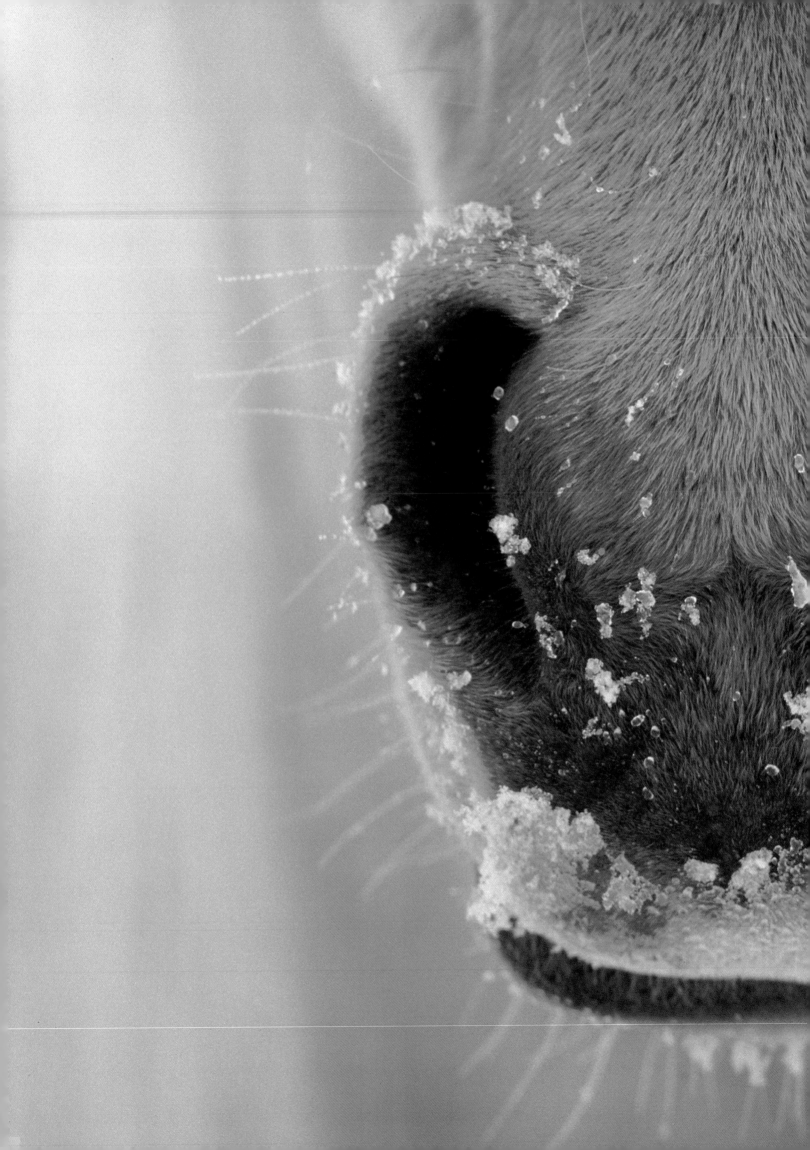

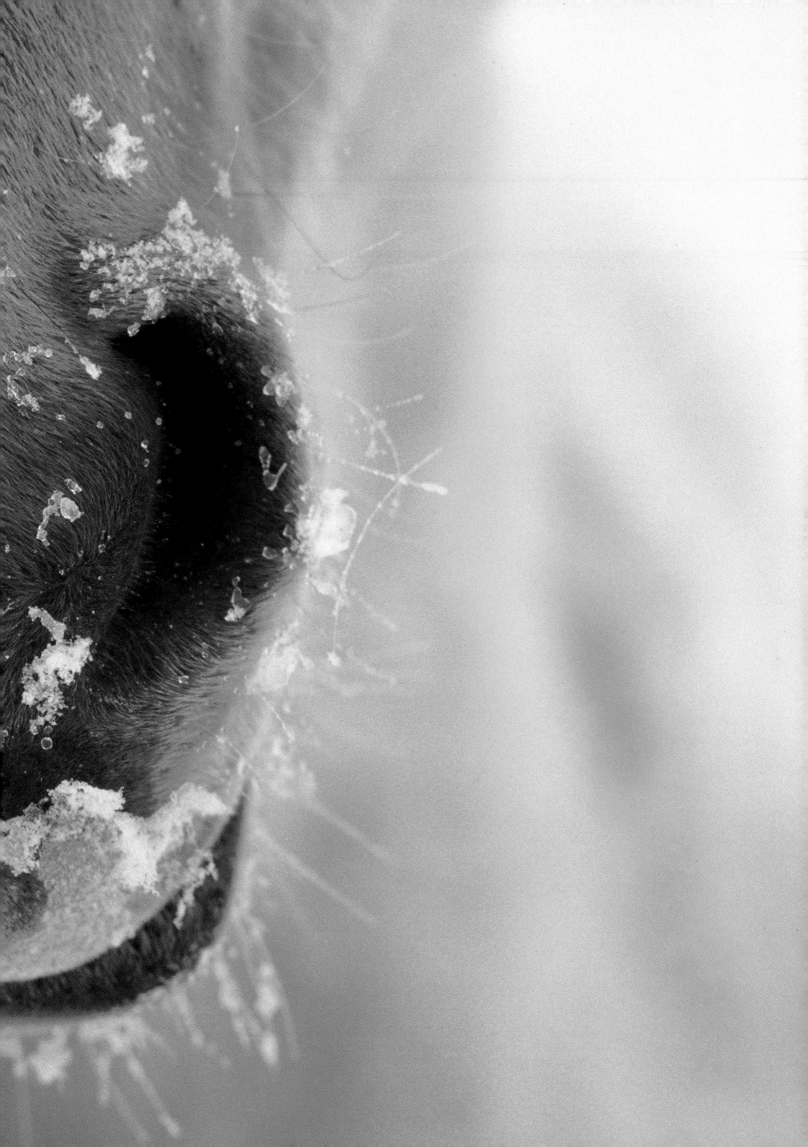

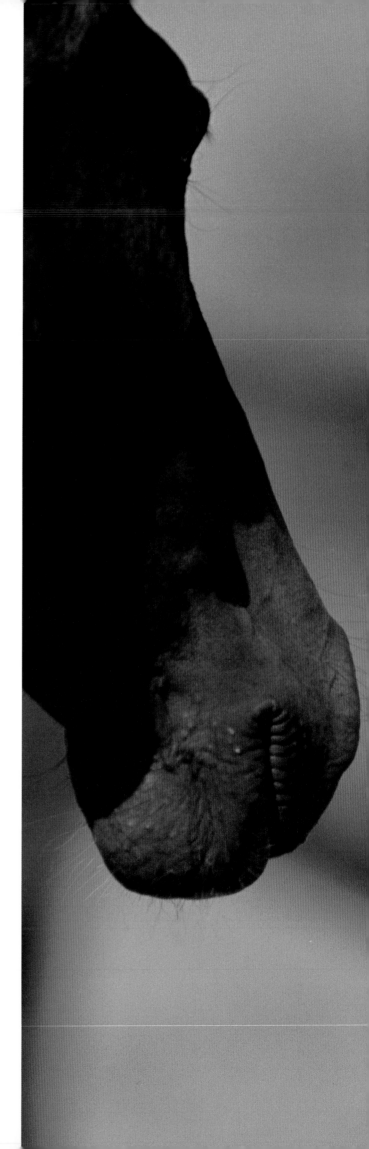

Language

D ANCING AND TOSSING HIS HEAD, the dark colt circles the stallion, his eyes wide and innocent. *Notice me, protect me, aren't I cute?* his demeanor signals. The colt is both deferential and demanding, and his father ignores him, turning aside, continuing to graze. *Leave me alone,* his body says. *Can't you see I'm busy?* The colt won't or can't give up. Finally, the stud turns and throws a solid shoulder block, pushing the colt back, knocking him off balance. Adult and baby stand in the grass. The colt turns away now, and the stallion stares sternly, ears flicking back and forth, body firmly planted, the boss.

The colt raises his head, bares his teeth, and clacks them together, signaling acceptance, baby privilege, subservience. Four times his size, the stallion shakes his head and moves away, his whole body communicating the universal impatience of a parent—*okay, this time, just don't bug me anymore, let me have some peace and quiet.* His son lowers his head to the patch of bunchgrass he wanted in the first place, placidly finishing the stallion's lunch.

Mustangs have a language involving complicated vocalization, a vocabulary of smells, and body signals. The tight protective structure of the group requires constant communication; the extremely social nature of the mustang makes them attentive. Individuals have distinctive voices and odors, instantly recognized and remembered by other members of the group. Status, sex, age, and character traits can be translated by an observer; within the band or family, a more complicated understanding of communication is highly developed. Mustangs know how to pay attention, and seem to be born with a need to interact with one another. In the wild, this intricate blend of language and need is one of the group's main survival skills. For example, a leader's command and warning of danger requires an immediate response; understanding protocol and hierarchy both in the band and on shared rangeland is knowledge communicated and followed. The education, protection, and nurturing of the young, and the intricacies of dominance and sex, have a system of signals and responses.

The triumphant challenge of a stallion's call and the soft whinny of a nursing mare are more than just hormones talking. They signal the integrity of the family, a language of cohesiveness.

Mustangs greet each other in a variety of ways—nose to nose, a casual how are you? accompanied by little snorts and nickers, as if they were indulging in small talk. A band constantly exchanges information; mares checking foals, complicated grooming rituals, pointing out choice forage, taking turns napping and watching the young. Within the tight little world of the mustang band, they have a language of support as well as information.

Mustangs have very extensive facial expressions. Mouths relax in loose lipped contentment, purse tight with fear or frustration. Nostrils flare with anger, or in an effort to gather more information from their acute sense of smell. They show pain and fatigue and curiosity, sexual interest and indifference. The sharp, pointed ears rotate constantly, acutely attuned to the sounds of each other and the environment. The ears also speak, sending messages: ears back means anger, warning; ears forward and erect, full alert; ears drooping, pain or boredom; ears moving back and forth, anxiety or irritation.

Every mustang reads other aspects of the language of body as well. Dominance and status, the basic hierarchy of the group, is maintained through body movement, from the snaking and herding behavior of the stallion, to the vigilance and leadership of the dominant mare.

They also constantly touch one another, communicating through pushes and shoves as well as tender mutual grooming and sniffing. Body language isn't just about control and dominance, although that is part of it; it is also about mutual support, bonding, maybe even something close to concern. Mares have a certain sisterly air within a harem band; since they form the strong nucleus of the family, they develop strong ties to one another. Rivalries may develop, tempers flare; but they seem to tolerate one another, seeking a sort of harmony by careful attentiveness.

It is this air of total awareness and physical intelligence that makes a mustang seem so ardently alive. Unlike many large grazing mammals, mustangs do not seem fear driven. They don't have the demeanor of victims or prey, like antelope or deer, although they are wary of humans. They live in supportive groups, have a well-developed system of communication, and seem to help each other under conditions of extraordinarily difficult terrain and weather. Mustangs survive as a group, as members of a unique tribe.

They move as one, this small band of mustangs. As if pulled by an invisible string, they all—including the foals—suddenly stop grazing, lift their heads and stare in my direction, ears pointed, nostrils trembling. I have been sitting here, braced against a rock, for two hours. Nothing has changed; I am downwind, partly hidden by brush, haven't moved or made a sound.

I glance around, wondering if some other predator has arrived; I know I have signaled no threat.

For hours, we've existed in peaceful quiet, a silent observer and a grazing band of mustangs. They are now on full, quivering, alert; clearly ready to run. I am irritated by their sudden, inexplicable spookiness. When they move, I can't follow; I've lost them. Why are they gathering, why is the lead mare abruptly wheeling and trotting through the brush, over the hill, leading them out of my life?

Standing, suddenly alone in the landscape, I wonder about the intelligence of mustangs. If they're so smart, why did they waste energy running away from someone who poses no threat? Why now, after several hours of mutual tolerance? I had hours of daylight left, and a comfortable perch; their departure is inconvenient. Or do I, possibly, feel insulted?

I shake myself—this goes too far, even for me, I think—and start back to my truck. I look forward to the hot coffee I am now free to drink, feeling bad about losing an afternoon of work, yet grateful to the mustangs for moving out.

IV

THE PRETTY LITTLE
PAINT FILLY

THE SUMMER I TURNED TEN WAS DRY AND DUSTY. The cottonwood trees lost their emerald glimmer early, and dead leaves and grasshoppers flattened under our feet. Over the badlands south of town and the prairie to the north, turkey vultures circled low, riding the carrion trails. Everything felt sucked dry, dogs panting in the shade, the river running low, the cool blue of Laramie Peak floating out of reach in waves of heat. Dust devils and tumble weeds were the only things moving, except the wind. The air was so dry it leached the moisture out of your mouth and left your eyes red and gritty.

We lived on the edge of a small Wyoming town, next to a dried up creek bed. My family had a few acres to support all our dreams—my mother wanted roses, my father tried a tree farm and planting wheat and raising chickens. Most of his schemes failed, but my mother managed to coax roses from the dust, blossoms as strange and foreign in the prairie landscape as our neighbor's plastic flamingos.

Sometimes the wind would carry a maddening hint, the smell of rain from some luckier place. Heat lightning and thunder mocked the ranchers as they hauled water or watched helplessly as livestock perished. Every Sunday, the preachers prayed for rain in front of congregations who believed hope was meaningless and only endurance counted.

My father came form a long line of Irish dreamers who never lost hope; he always believed the deal he was making was the deal of a lifetime. He bartered and traded for piano lessons, and meat, and coal for the furnace. Instead of money, we had the currency of the unexpected deal, like the purple Kaiser car and the pregnant rabbits and the twenty gallons of exploding beer.

But that summer my father made a deal that left me mute and trembling, so happy I spun in my tracks until I fell in the dust. An old friend of his, a horse wrangler, needed a place to stay. Shorty had three horses he was green breaking; and was trading one for a roof and pasture. That was my birthday present.

When Bingo arrived I was too excited to even see her, at first. I knew she was a paint filly, a mustang from the Red Desert of southwestern Wyoming. Shorty was a mustanger, part of a crew who helped catch and ship horses for slaughter. The best and healthiest were sometimes saved, to be sold as cowponies or rodeo stock. I knew this the way a child knows things, that adults do bad things it's best to ignore. I was ten years old, about to get my very own filly, and I didn't care how, or where she came from.

I simply worshiped horses, loved everything about them, from their strangely naked heads and dancing feet to the richness of their smell. Around horses I became fearless, and from the time I could walk would run after any horse, climbing over and under and around them. I spent hours running in the prairie, tossing my hair, lifting my feet high, creating an imaginary wild horse band as playmates. I thought I could speak horse; most nights, and even in daylight, I dreamed I was a horse.

Bingo came filthy, bone-thin, sore-footed, hungry and thirsty; she could barely stand, was terrified and near death. Now I know something about her capture: harried by aircraft, bunched together and chased until horses were dropping; crowded into some kind of fenced enclosure, she had lost everything familiar. Her family and her freedom and the places she knew, all gone. Some didn't even survive the chase; others were killed because they were too old, or too wild, or maybe there was no room on the truck. Bingo was one of the lucky survivors; at least, she wasn't dog food. At the time, I just saw a horse that was mine.

Because Bingo was in such rotten shape, she let me handle her from the beginning. Shorty wanted to keep her snubbed to a post, tied with a short rope so she couldn't move or kick; I insisted on using just a halter and a lead. I wanted her to come to me, learn to love me, accept her new home without coercion. At first, my way seemed to work—she was too exhausted to fight me. I spent hours cleaning, brushing, trimming, combing her until I knew her body better than my own. I hauled countless buckets of water, pulled clover from the creek bank, begged grass clippings from the neighbors; I even stole all the baby carrots from someone's garden as a treat for her.

Physically, Bingo healed; as Shorty said, she sure cleaned up pretty. She was a deep red-brown and white paint, with both colors in her mane and tail. Her head was fine and delicate, showing Arabian ancestry, and she had the neat body of a small Quarter Horse. One blue eye gave her a slightly startled expression. I imagined Bingo was the direct descendant of the first mustang, a spotted Spanish horse from some 16th century expedition.

Bingo became stronger and declared war on every human being except me. She refused to let a rope or a blanket or a saddle anywhere near her; would come when I called, but attacked Shorty with her teeth bared and feet flashing. Sometimes she would even bite or strike out at me; and I would stand, heartbroken, behind the gate and watch her pace until she was exhausted. I thought I understood. After all, all my life I wanted to run free with the wild mustangs, too.

Shorty told my parents that Bingo was useless, a renegade who refused to be civilized. I was a country kid, and I knew that no one would give room to a creature who didn't work, not a person or a dog or a pretty little paint filly. But I was in love, and ignored the warning rumbles from the adults. The rest of that summer we were all as agitated as the dust, going back and forth, unsettled and clueless. Shorty took the failure personally, a horse he couldn't train;

my father dismissed Bingo as a deal that just didn't work. My mother understood a little—there were those roses—but she was afraid Bingo would hurt me. They didn't see Bingo, they just saw an animal who refused to conform.

Of course, the end was inevitable, and sudden. Summer ended, the Wyoming State Fair and Rodeo was in town, and my father sold Bingo to a bucking string when I wasn't looking. One minute she was there, head up and pacing the dust, homesick and untamed; the next time I looked, the pasture was empty and the dust still. There was no use crying and screaming, although I did; a deal was a deal.

I still remember the way she smelled, and the feel of her body quivering under my hand. I loved her wildness, and her dainty feet, and the one blue eye. What happened to her was cruel, and broke my heart. I was only ten and couldn't save her. But sometimes, when I am out in the Wyoming hills, I close my eyes and imagine a pretty little paint filly running with the wind.

V

PARTNERSHIP

⚬

THE WRANGLERS STAND IN THE WEAK SPRING SUNLIGHT, exchanging old and quiet jokes, knocking the mud off well-worn boots, pulling at their caps and thick belts. Near the pens, a few seekers clutch new halters and braided cotton ropes, scanning the posted lists and whispering like conspirators. Spring green meadows roll behind white running board fences and the barns, fresh painted gray and white, are slick from the morning showers; neatly tilled soil waits for flower beds around the gravel paths.

This is Virginia, home to Thoroughbreds and fox hunters, elegant dressage horses and well-groomed ponies. But for the next few days, the Virginia Horse Center is sheltering an entirely different sort of horse—the mustang. Over a hundred mustangs are contained in portable pens, brought here from the Nevada foothills and Rocky Mountains by the Bureau of Land Management, a thousand miles from the smell of sage and juniper. They have been gathered and fed and doctored and shipped here to Virginia, unloaded and separated and tagged, mustangs selected out of free roaming herds and offered for adoption.

The mustangs are confused. No longer part of a band, the subtle and complicated patterns of behavior they know mean nothing here. Separated according to sex and age, they are not members of an extended family but individuals, an entirely new way of being. The code of conduct governing eating, drinking, and responsibility disappears when the supportive structure of the wild band is gone. Naturally herd animals, mustangs become less distinct, oddly diminished when forced to become autonomous. Their identity changes. They occupy space in a different way, and react to other mustangs with less recognition, less concern. It's not just fear, but something we would call alienation in a human being; an animal out of place and

estranged. Strange smells, strange diet, strange light, everything new and unknown. For these mustangs, it must be like being picked up by a tornado and deposited on Mars. The miracle is not in the force of the tornado, but in the amazing resiliency of the survivors.

And in fact they are in relatively good shape. Of the 120 or so offered for adoption, only two show obvious signs of stress or ill health. The rest simply don't know how to behave without the structure of the band. They look like mustangs; they act like prisoners.

The mustangs mill and raise dust in the crowded pens. Prospective adopters walk slowly, mesmerized and enchanted by the muscle and bone and spirit of the animals. This is a big decision, a big responsibility, for both the people and the mustangs. The adoption fees are small ($125) but the commitment, in terms of time and money, is enormous. The faces and voices of the people are full of hope and determination. Adopting a wild horse—and some of these people have already driven hundreds of miles with nothing but a wish and a dream—is a chance to partner a uniquely American animal.

An older couple tells me they've already adopted two burros, and now want a mustang running free on their ten acres. They don't plan to ride the animal; they want to save a life. Many people seemed to have similar motives; they weren't necessarily looking for a serviceable horse, they simply had an impulse to preserve what they regard as a living piece of American history. The wranglers and Bureau of Land Management staff offer careful information, and discuss the law, the care and feeding of mustangs, even have several well-trained animals on display; but this crowd is moved more by the image of a wild mustang. What happens when they get it home, nobody can predict.

At adoption events in the West, many people look for a child's first horse or a good, solid working mount. They know the type of mustang they want for a particular purpose. In the East, choices seem to be made from the heart. Conformation, age, color, and sex aren't as important as the simple desire to know a mustang. Selections are made quickly—people draw numbers, and can choose two prospects—and seem impulsive. I just like that one, they say. He's all by himself, or looked me in the eye or seems like he needs me.

The adopters are as shy as the mustangs. They speak softly and move slowly down the row of pens. The mustangs respond with varying degrees of alarm, aggression, fear, and curiosity. Studs crash against the metal bars, snap at each other, cut the air with shrill challenges as if asking who is in charge. Mares bunch in corners, the shyest seeking to hide, the largest shifting position and presenting powerful hindquarters, as if they were defending themselves against a pack of wolves. A group of yearlings, all legs and rolling eyes, move as one, as far away from us as they can manage.

It is strange, I suddenly realize: mustangs in the wild constantly touch each other, standing shoulder to shoulder in all the open spaces; here, thrown among strange mustangs gathered from different bands, it is as if any physical contact makes them crazy. Like strangers everywhere, they find no comfort in forced proximity. The touch they desire is no longer available. Crowded in groups, they remain unnaturally and visibly alone.

Today, a steady Virginia rain softens the air, spring weather mustangs never saw in Nevada or Wyoming or Montana, where the arid landscape makes do on less than twelve inches of rain a year. They eat good quality hay, orchard grass and timothy, pale green and nutritious. The wranglers keep water and hay available, causing small, contained stampedes in each pen. This is a day many mustangs will find a home, a new environment, the beginning of a new life. They are already learning fast, drinking from a barrel, eating hay, living without expectations or understanding.

And as much as a romantic and impractical heart wishes it weren't so; as fervently as one believes that mustangs belong to the land and the land to them, still, the adoption program is a way of ensuring survival for all mustangs, and with them, the unique mustang genetic heritage.

They do look weird, however, next to manicured paddocks and carefully planned arenas. They even have shelter, a roof over their heads. This is a better fate than mustangs could

expect before 1971, when their population of several million diminished to less than 17,000 mustangs hiding in the western United States.

Public awareness, fired by the spirited advocacy of people like Velma Johnson (Wild Horse Annie) of Nevada, finally forced political action. Congress passed the "Wild Free-Roaming Horse and Burro Act" in 1971, declaring that mustangs are "living symbols of the historic and pioneer spirit of the West." This law, and subsequent amendments, was meant to ensure mustangs perpetual recognition and protection. The federal managers of the public lands where mustangs still survived were charged with the development of programs to protect the animals and their habitat, and implementing programs to address competing demands on public land use.

Mustangs survived blizzards, drought, parasites, predators, wars, the industrial revolution, and diminishing habitat. With the passage of the 1971 Act, they entered a new arena, wild gladiators fighting for public attention and competing for limited resources on public land. The Bureau of Land Management is the referee, setting the rules "in accordance with the principles of multiple-use management." Mustangs, stock growers, land developers, wildlife specialists,

plant biologists, mineral companies, water users, state and federal officials, are all jockeying for the public space—the allocation of resources—on public land.

We've seen wild horses extinct once before on this continent. And we can no longer claim ignorance about the long range impact on fragile ecosystems when the balance tips too far in favor of any single use. Federal managers have a Solomonic task in assessing the value of competing claims on our public land, including the soil, the plant and animal life, and human economic exploitation.

Initially, mustang populations protected by law from human predation and commercial exploitation proved only too successful and the herds increased rapidly. The Adopt-A-Horse Program began in 1973, when 23 horses from Montana were gathered and sent to new homes. By 1976, The Bureau of Land Management began using helicopters, both to monitor herds and help during round-ups. The size and complexity of the effort to implement the law keeps increasing. Research projects, both on the unique characteristics and genetic inheritance and on the more pragmatic problems of population control through more selective removal and contraception, are now mandated by amendments to the original law.

Several states have instituted experimental prison projects, where inmates gentle carefully selected mustangs for prospective adopters. The union of man and mustang, helping one another adjust to a different discipline and a different life, provides a public service to the mustang, the new owners, the inmates, and the prison system.

Today, several refuges for mustangs exist, in Nevada, the Pryor Mountains on the Wyoming-Montana border, Little Bookcliffs in Colorado, and the Kiger Plateau in Oregon. In South Dakota, a private sanctuary, The Institute of Range and the American Mustang (IRAM) has been operating since 1988.

Since 1971, these efforts and others have struggled to increase awareness, understanding, and protection for the mustang. Like the creatures themselves, mustang conservation plans are constantly changing and evolving, responding to the shifting winds of political expediency and pragmatic discovery. No one knows the final fate of the mustang; the bare outlines of plans and hopes and scientific information, and an impulse toward preservation, are all we can build.

One reason mustang populations increase is because virtually all the large predators are gone. One player in nature's balance, the culling of herds of weak and sick animals by predation, has been eliminated by cattle and sheep ranchers to ensure the safety of calves and lambs. The alternative to killing 'excess' mustangs (by man or the reintroduction of predators) is to keep mustang populations at a level posing no perceived threat to habitat or human economic interest.

That is how 120 or so scruffy and uncivilized western mustangs landed here in Virginia's horse country. Almost half of the mustang adoptions occur in the eastern United States, a new migratory pattern that makes a strange sort of sense. People move west, need more land and water and game animals and power; mustangs are pulled out of the deserts and mountains and canyons and taken east to new homes.

Three of the pens hold male mustangs, mostly studs, a few geldings. They are dark horses, black and brown and bay, probably from Nevada, where 65 percent of the horses carry those colors. One animal really worries me: he is listed as 14 years old, newly gelded, and the possessor of a head curved like a Shriner's sword, a nose Roman and noble and containing more bone than any horse could reasonably carry. Over the next few days, I find myself watching him, hoping someone will give this old fellow a home, but I can't imagine why anyone would

165
PARTNERSHIP

want him. He isn't pretty, he's been gelded as an adult, and his behavior in the pen is irritable, nasty, and intractable. This one will require a special person.

Yet here, as at other adoptions, I see people find special mustangs. They want inexpensive solid horses or intelligent trail horses or affectionate family pets or trainable gaited horses and a mustang can be picked who will fulfill any of these—and more—human needs. Here, in the Bureau of Land Management's adoption program, in a very unlikely setting, I see another example of the amazing adaptability of the mustang—they can seduce humans into becoming their willing partners and protectors.

Although removal and adoption remains the main management tool of BLM efforts to maintain mustang population levels, by 1981 it was clear that other methods, including sanctuaries and contraception, needed to be explored. There is now a fertility control program, which combined with more careful selection of mustangs for adoption, will keep birth levels down. The hope is that the contraceptive method used will stop selected mares from conceiving without causing disruption in the delicate social behavior of wild mustangs or resulting in irreversible depletion of their genetic heritage.

If left to nature, but without natural predators, expanding mustang herds would damage the fragile and unforgiving western rangelands. Mistakes are not easily reversible in arid and semi-arid country: the century-old scars of migratory trails have never healed; the Oregon Trail, the Overland Trail, the Mormon Trail still cut across the plains. On some parts of the range, overgrazing by domestic livestock has degraded the native grasses and forbs to the point of extinction, and some areas struggle against aggressive invasion by non-native species like cheatgrass. Even in the late 1800's, the high plains supported large herds of wildlife and domestic animals in numbers that would not be possible today.

The destruction of habitat by our species can be especially devastating in semi-arid landscapes like the American West. To the human eye, the high plains can seem stark and empty; yet on the ground, it holds a rich and fragile ecosystem. Reverence and respect for the landscape are beginning to replace opportunistic greed as our understanding and knowledge increases.

The union between land and mustang and man is also an educational opportunity. The federal managers, biologists, behavioral scientists, fertility specialists, and mustang protection societies are engaged in a constant dialogue, searching for a way to celebrate and maintain the special mustang character and rich gene pool; and to balance that desire with competing human demands on public land. In the years since the passage of the 1971 Act recognizing the mustang as a unique American animal, resolution remains elusive. Hunters want more allotments for bighorn sheep, elk, and deer; developers want more roads, coal companies want open mines, farmers want water, stock growers want the grass; none of these groups favor the mustang. And the BLM must maintain a balance between them all.

Yet mustangs continue to engage the concern and protection of the American public. Adoptions are well attended, and most of the gathered animals find homes. Tourists are beginning to seek out places where they can view mustangs in their natural habitat. Public pressure on Congress has so far maintained the law protecting mustangs. But people must work hard to continue to preserve the mustang's special status. Already the Red Desert is threatened with 800 exploratory gas wells, and part of Nevada's wild habitat is a weapons-testing range. In the Pryor Mountains the introduction of greater numbers of bighorn sheep for trophy hunting could mean eventual reduction in the mustang herds.

On the last day of the adoptions in Virginia, one of the wranglers told me that the old Roman-nosed gelding was on his way to the next adoption center, somewhere in Georgia. I hated the thought—it seemed somehow undignified that a stud who had survived 14 years in the tough Nevada rangeland should first be castrated, and then toured around the adoption circuit, a wandering, unwanted supplicant, without home or family.

On the other hand, before the wrangler loaded the last lucky, adopted mustang, he said that he'd been thinking; maybe he had a little room left at his place, and there certainly was something about the old guy that grew on you. I wanted to dance, and kiss them both, and imagined the two old guys sitting somewhere in the sun, telling each other stories about the wild mustangs they had known.

GATHERING

MUSTANG

🐎

AROUND THE WEATHERED WOOD CORRAL, the ground is packed hard and sterile, made barren by the flying hooves of thousands of terrified animals. No grass grows, no insect moves, no snake leaves a trail. A tumbleweed moves across the gray earth and finds no purchase, not even a rock, and blows on down the trail. The walls of the small box canyon are eroded by time and wind, sagebrush clinging to the few patches of dirt, a clump of thistles standing on top, purple heads against the sky.

You can trace the path once followed, back up the draw and onto the plains. Your eyes can see dead ground twisting through eroded walls, sandstone worn smooth by passing bodies. Once, this place may have held water, cut by an ancient flashflood. Or it may have been a soft spot in the landscape, where a receding glacier left sand and dirt. Thousands of years of wind have carved a shape, a secret hole in the ground. Plants, insects, burrowing animals all sheltered here. It wasn't always a dead spot in the landscape.

Man must have brought the wood used to fashion the fence, gnarly cedar posts and odd lots of lumber held together with wire. Not all the pieces are here—but it is still recognizably a fence anchored in the earth. Clinging to pieces of wood and buried in the ground are bits of bleached canvas and burlap. Once, sixty years ago, the rough fabric and gray, splintered wood formed a trap hidden in the folds of earth, shaped like a bottle, the neck wide open now but the bottom still holding. The closed end backs into a twelve or fifteen foot steep slope, studded with rocks and the stunted shapes of sagebrush and snakeweed.

An old cowboy, a mustanger, told me about this place. I've been lost most of today, following paths that lead nowhere, looking for vague landmarks that seem to disappear as I approach.

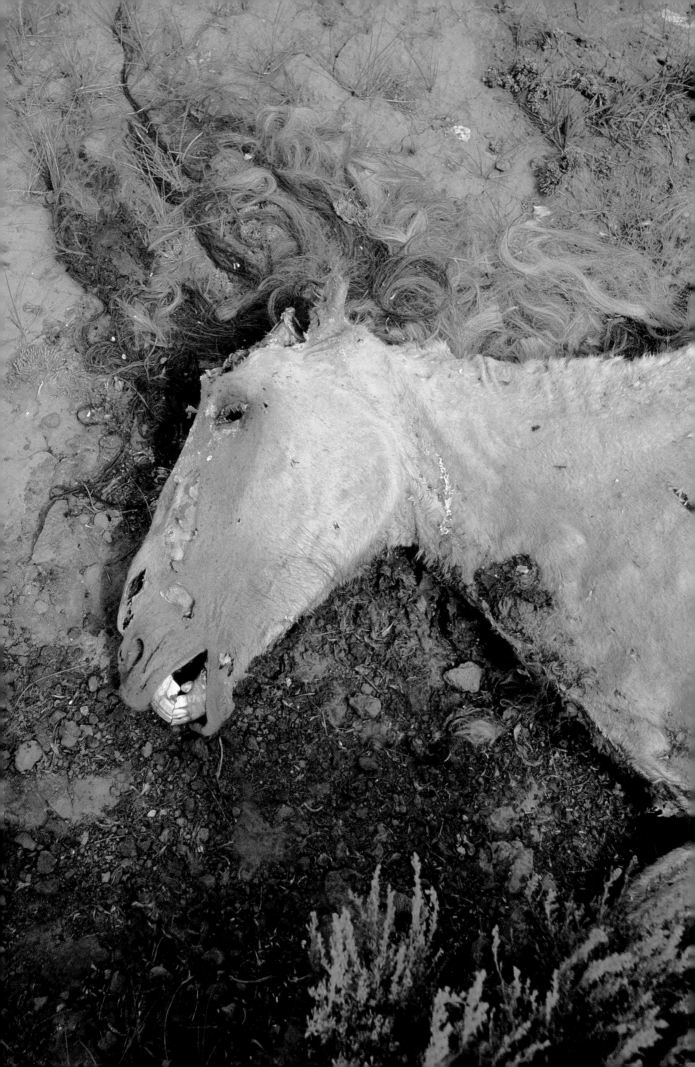

Some dirt roads cut straight across the landscape, ignoring shape and contour, drawn by government surveyors. Meandering ruts, which began as old game trails, start and stop without warning. Deep circular depressions, filled with drying grass and rustling wild sunflowers, mark ancient buffalo wallows. Sagebrush, prickly pear, cheatgrass, needle and thread grass; even this hard-grazed landscape shelters a variety of plant life. No trees, no signposts, no sign of water. I've seen antelope, two coyotes, crows and turkey vultures. I've heard nothing but the wind and my own vehicle struggling across the world.

West of the big flat top, across a dry crick bed, he said, *didn't use to be fences after the first turnoff. Look for two piles of rock about a half mile apart; one kinda looks like a bench. The pen was just a little to the south, through those rocks.*

The boss and supply truck would come first, setting up camp near the rocks. The mustangers rode out, scattered across the prairie, waiting for the airplane. When they heard the motor straining, filling the air, they just rode toward the dust. The pilot would gather the scattered bands of mustangs into one large, running herd and bring them to the cowboys.

All we had to do was pick up the stampede and turn them into the draw, he remembered. *Those broomtails would go anywhere to get away from that airplane, and before they knew it, they were trapped. Run them into the catchpen, close the gate, that was it.*

Not quite—some lead mares and stallions would sense the trap and break out of the herd, taking others with them. Exhausted animals would pull up lame, or go back for a foal. The men learned to kill any stragglers, any mustang that broke away from the herd. The rest would bunch close together, crazed by gunshots and the smell of blood and the screams of the dying. You had to keep them moving, running into the trap, or they'd scatter, get away.

I find the rocks—more a collection of table shapes than a bench, I think—and climb up on higher ground, trying to spot the old catchpen. Finally, it is the dead ground that stops my eye, a place where the dirt was packed hard by running, terrified mustangs 60 to 70 years ago and still nothing grows. Two faint parallel tracks run off this hill; the trucks were up here.

They wanted all the wild mustangs off this grass, grass they needed for cattle. The mustangs belonged to nobody and were of no more use than the prairie dogs the men gassed, or the coyotes they poisoned or the antelope they shot.

Since 1804, when Zebulon Pike first saw the western grassland, men dreamed of great herds of docile cattle converting grass to money on the hoof. The wild herbivores were slaughtered; the buffalo eliminated, antelope, deer, elk butchered. Predators—wolves, coyotes, mountain lion, bobcats—were tracked and killed. The dream included nothing but domestic livestock. And finally, men went after the wild mustangs. In 1930, almost 30 million pounds of horsemeat came out of the West, most canned as pet food. The profit was secondary to the intent: preserve the grass for beef and sheep, get rid of the mustangs.

The wind comes down the walls of the trap, rattling the bones of the old catchpen. It sounds like hooves scrambling, trying to climb the canyon walls. I can feel the terror hammered into the ground, pushing all the life out of the soil. How many mustangs began to die here; how many never left the canyon? Hundreds? Thousands?

The old man said he didn't know. *The Bible tells us we're put here by God to subdue the land and the animals,* he said, *and by God, we did it!*

The idea that man is only a part of the earth, not the master, didn't have much relevance for people struggling to survive, especially on the frontier. The thought that we can modify and alter the juggernaut of progress in positive ways, that we can change our attitudes toward the earth, is relatively new. The thought is both romantic and radical. But so, ultimately, is the national character shaped by the frontier experience. Optimism, toughness, ambition and an impulse toward conservation are qualities valued in the wilderness, revered in the American self-image. Reverence toward the earth, concern for all life, and a curiosity about the complex relationships that compose our environment nourish the roots of thoughtful preservation and, more importantly, enhance the strong need human beings have for a sense of place, a natural involvement with the landscape they occupy. Now we are beginning to see the ways to change the way we move on the earth.

As we moved west, not searching for our niche so much as assuming we owned the landscape, we altered the ecosystem with the force of a meteor, destroying species and changing the land forever. But we also began to appreciate what we saw, and we were moved to preserve the unique places of a vast country.

Much of the world has become totally civilized, bereft of wild places and wild things. Here, in the United States, the enormous western landscape afforded an opportunity to prevent that loss. National parks and forests, and wildlife sanctuaries, were part of our early public effort to conserve life. By 1916, there were thirteen national parks, created for public use and enjoyment.

Now, we have moved beyond the utilitarian idea of preserving open spaces solely for human use. We are beginning to understand a need to conserve the entire web of life, nature in all its forms. The idea of sanctuaries, areas where the earth's entire ecosystem would be protected, was politically expressed by the passage of the Wilderness Act in 1964.

Consider how far we have come, given the amount of knowledge we have and the way our continent was settled. We have learned that our actions have consequences—even to the air we breathe, the water we drink, the food we eat, the earth we stand on, and the other forms of life on the planet. We are learning the ways things connect, and that death lies in ignoring connections.

Mustangs are part of our connection with the landscape and our own history. When Congress declared them living legends, its members were vaguely thinking of the Horse Culture and the settling of the West. That is true; mustangs were a necessary player in that part

of our history. But the mustangs are more than that, almost more than a legend or a horse. The circle is so clear—the mustang returned twice from extinction at the hands of man. It is the cycle of regeneration, rebirth, redemption.

Mustangs are continuing to evolve in the wild places, adapting and surviving and thriving. It is as if they are bringing us a message in their being—a lesson in how to live.

In order to survive, mustangs have become more frugal, able to do more with less. Their bodies are pared down to tough muscle and bone, hardy and strong. They live in close social groups, interdependent and supportive. They are protective and sensitive to each other, and are not territorial. Their world is highly sensual, rich with meaning, responsive to the earth and seasons. In order to survive, they have made us a partner. Protected by law, they are public animals on public land. And they exist as an integral part of the landscape—making it more beautiful.

Earth. Sky. Horse. Man.

Wildness

VII

When the wind comes out of the west and the sky is a blood red bowl over the empty landscape, you can hear them. Coyote, bobcat, wolf, mountain lion, bear, mustang, prairie toad, ferret, big horn sheep: their wildness echoes across mountains and grassland, in hidden canyons and down ancient dusty trails.

The wildness speaks directly to the human soul, evoking knowledge and truth and fear so old we have forgotten the beginnings. The vision of open spaces, the freedom and beauty of wild creatures, the purity and exuberant expansiveness of the landscape is a challenge to all of us, a message carried on the wind—in this natural world, what will endure? The predators hunt, the prey develop survival skills, the plants adapt to the weather. Everything in nature moves and changes; shape-shifting mountains and rocks, drifting dust and blowing seeds, soaring raptors and agile antelope. Nature requires movement. Balance and harmony exist only in tiny increments of time in nature, moments of peace and stillness where things hang perfectly formed—the birth of a fawn, the opening of a mariposa lily or the song of a meadowlark. As they occur, they are perfect and whole. Yet they are part of a constantly changing mosaic; the fawn will grow into a deer, the lily will become a seed, the song will fade into the wind.

The most powerful agent of change, and the most potent threat to wildness, is now mankind. We arrived on the North American continent some 10,000 years ago, practically yesterday in geological time. Our hand in shaping the ecology, the kinds of plants and animals that still survive, even the weather, has been a force beyond description. Human beings arrived and within a few thousand years thirty-nine large mammals disappeared from the continent. Gone were the horse, camels, the sloth, the mastodon, and the mammoth. The extinction of these animals was just the first blow in our long history as a catastrophic species, and we did so—stripped bare one part of nature's inventory—armed only with the rudimentary technology of the spear.

Over time, our technology and ability to shape and control our world became the greatest power for change since the planet was formed. We began as hunter-gatherers and became a people who would domesticate animals, the landscape, and fire; who would build cities and cultures and religions. Having done all this, we stand on a planet changed forever by the heavy hand of man, the Earth at risk from unimaginable pollution and the diminishing returns of exploitation of great natural resources. The Earth itself is in a state of crisis, providing a window of opportunity for human beings to become agents of change in a new direction.

The opportunity for responsible and aware stewardship is especially urgent and possible in the American West. Over half the land in eleven western states is owned by the American people, public lands administered and managed by the federal government. Hand in hand with the greed and fear that led us to destroy entire species and poison ecosystems is a sometimes unacknowledged love of nature. Buried somewhere in our memory is a sense of our wild self in nature, bare feet touching the earth, face turned toward the sun, body moving in the wind. As a nation, we have a complicated and contradictory relationship with the landscape we inhabit: notions like 'Manifest Destiny' allowed us to move through the territory in a mindless feeding frenzy, using irreplaceable resources with no thought for the future. At the same time, the sheer beauty and immensity of the country created a movement toward conservation and preservation, setting aside land for National Parks, Forests, and Grasslands; passing legislation to preserve and protect both habitat and wildlife. The concept of the village green, public space held in common trust, has been expanded to protect millions of acres and preserve wildlife.

In our open spaces, in the land and the wildlife and the forests, the rivers and mountains, lives the American spirit. Freedom, openness, respect for all life, curiosity, flexibility, tolerance; these are values that shaped our national character and allowed a democracy room to develop and endure.

The contradictions and controversy, chaos and rigidity, greed and sentiment, that inform human endeavor are especially obvious in the political arena where battles over public land are fought. Nature cares nothing for politics, doesn't recognize economic requirements or reward only the prudent and well intentioned. She is a difficult partner, in life and in politics. Coyotes hunt poodles, cougars stalk joggers, wolves eat lambs; no one, least of all Congress, finds it easy to cope with these facts, any more than they can prevent a killer blizzard or the eruption of a volcano.

Yet our common awareness of ourselves as a creature of wildness, as a member of the natural community and part of the ecosystem has always been there. Prehistoric drawings of animals and plants; the beginnings of culture and religion; the need for love and family; these are all rooted in the natural order of things. In order to survive and endure as a species, we developed the power to destroy, from the technology of the spear to pesticides and herbicides. At the same time we were developing an intelligent consciousness that both informed us life was harsh, brutal, and short and gave us the ability to appreciate and even worship life in all its wondrous diversity.

Think about it—the same naked primate who cowered in fear from the bear, developed technology to defend himself and left a record showing intelligent appreciation of the beauty, fierceness, and power of the beast. Now we are moving beyond that primitive consciousness, beyond fear, through awareness and appreciation of wildness, into a recognition of our particular responsibility to preserve and defend the remaining natural landscapes and the creatures with whom we share the Earth.

This is the measure of how far we have come in our own evolution. From being so fearful and hungry that we changed entire ecosystems and caused the extinction of many forms of plant and animal life, we are moving to an acceptance of our responsibility to preserve life in the wild.

This book is part of a series—American Wildlife in American Spaces—which celebrates that partnership. Each title will seek to widen our experience and understanding of particular creatures and particular habitats. They will speak directly to the spirit of America, a spirit that is not afraid to treasure wildness and nurture freedom.

Most of us will never hear the howl of a wolf cutting through a winter night or see a mustang stallion silhouetted against the dawn, but our national soul needs to know these things exist, and the body politic needs to take a hand in their preservation. Our imaginations need these free and wild creatures living in open spaces, to keep ourselves dreaming and living free.

GLOSSARY

Adopt-A-Horse Program: For more information write: Bureau of Land Management, 411 Briarwood Drive, Suite 404, Jackson, Mississippi 39206. Tel (601) 977-5430.

Andalusian: The name given to Spanish bred horses, chief descendants of the Barb.

Appaloosa: Selectively bred for spotted coat pattern by the Nez Perce: generally larger than other Indian horses and now a recognized breed.

Arabian: One of the oldest breeds, regarded as the foundation stock of Thoroughbreds. Noted for speed and stamina; characterized by a fine, concave head with small muzzle and large, expressive eyes.

Bachelor Band: A temporary grouping of young stallions, who have been exiled from the harem band after reaching sexual maturity. Some older, unsuccessful stallions also seek out bachelor bands.

Barb: North African horse, a major foundation breed in Spanish horses and the Mustang. Spare and hardy, with a long, straight or convex head.

Bay: A reddish brown coat color with black points (mane, tail, legs).

Breed: Category of horse sharing characteristic size, color, conformation denoting a common ancestry. Domestic breeds are selectively bred long enough to ensure consistent production of defined stock, and are registered by a governing society.

Bunchgrass: A hardy, native perennial of the semi-arid grasslands. The dominant grasses like grama, little blue stem, and buffalo grass are nutritious forage plants, adapted to dryness with large root systems. In some areas, over-grazing has destroyed the native grasses.

Cheatgrass (Brome): An opportunistic and imported annual grass, usually considered inferior forage, which has replaced native grasses throughout the West.

Colt: A young male horse, from less than a year until sexual maturity.

Conformation: The size, shape, and proportions of a horse, a specific way of describing the animal.

Cow Hocked: Refers to hind legs which turn in, like a cow's, instead of straight when viewed from the rear. Frequently seen among mustangs, may serve some adaptive purpose.

Dorsal Stripe: A stripe of dark hair, from neck to tail, regarded as part of the primitive color pattern of the horse.

Draft Horse: One of the descendants of heavy northern horses, whose broad, thick proportions allow for exertion of great strength. Examples are Clydesdale, Belgian, etc.

Dun: Color pattern found frequently among mustangs, usually marked by dorsal stripe, black points, and sometimes zebra stripes on lower legs. Basic colors vary from Buckskin (yellow or tan); Grulla (bluish gray); to red.

Ecosystem: The system of linked animals and plants that have evolved together in a certain environment; usually, all elements of an ecosystem are mutually dependent for survival.

Eohippus: 'The Dawn Horse', the evolutionary ancestor of the modern horse, a small browsing animal which existed sixty million years ago in North America.

Equus: The modern horse (*Equus caballus*) evolved in recognizable form one million years ago. Migrated from North America to all the world's grasslands and developed into many different breeds and types.

Filly: A young female horse, from less than a year until sexual maturity.

Flehman: Part of an olfactory or smelling response; involves stretching the neck, raising the head, and rolling the lips back toward the nostrils. Displayed by stallions during breeding season.

Foal: A young horse, from birth to usually one year of age.

Forbs: Perennial herbs with broader leaves than grasses; drought resistant, these include sunflowers, goldenrod, loco weed, and clover.

Gaits: The natural paces or speed and pattern of movement; the four gaits of a horse are walk, trot, canter and gallop.

Gelding: A castrated male horse, surgery usually done before sexual maturity as a method of population control or to gentle a horse colt.

Gene Pool: Refers to the genetic heritage of a group or individual; the governing code of life, mapping the basic structure of an organism.

Grooming: One of the social behaviors of the mustangs, when members of a band stand and, using their teeth, clean and rub one another. Also used as courtship display by stallions and as part of bonding ritual between mare and new born foal.

Grulla: A 'primitive' or dun color, dark or light blue-gray with black points. The mustang bands, like those in the Pryor Mountains, which have many grulla horses, are thought to be closer to the original Spanish horses. Grulla coloring usually includes black points, dorsal stripe, and zebra markings on legs.

Habitat: The place or community where a plant or animal lives and grows.

Harem Band: Mustang family group, consisting of a dominant stallion, mares, and juveniles. The natural and basic social unit of the mustang.

Height/Hands: Medieval unit of measurement based on the width of a man's hand; equals four inches. Horse height is measured from withers (the highest point of the back) to the ground. Mustangs can vary in size but average 14 hands high.

Herd: A large group of grazing animals who occupy the same habitat. Mustangs form herds occasionally, usually when under pressure because of weather or when forced to by crowded conditions. Mustangs are not territorial, but nomadic, and are generally found in family units known as bands. Migratory animals, such as elk, form large herds only when moving from mountains to valleys in the fall.

Horse Culture: Refers to the brief period of American history from the 1500's to 1900 when the horse was the preferred transport. In the west especially, the mounted Indians, cowboys, soldiers and settlers live in popular imagination.

Longevity: Mustang life spans vary, but rarely exceed 20 years. The average life span of domestic horses is 25 years.

Mare: A fully mature female horse, capable of reproduction, about age four and over.

Mestena: Spanish for stray or ownerless beasts; became "mustang", the small hardy horse of the plains, foundation of the Horse Culture.

Muzzle: The soft mouth and nose of a horse; the size and shape of the muzzle can suggest ancestry.

Paints or Pinto: A two colored horse, white and black or brown. Among the Plains Indians of the 19th century paints were prized for color and hardiness. This is a color description and not a horse breed, although Pinto and Paint associations maintain a registry. Thought to be one of the original colors of the 16th century Spanish horses.

Palomino: A gold colored horse with a white mane and tail, a coloring found in many American breeds, particularly Quarter Horses. The American Palomino probably derives from the original 16th century Spanish imports.

Quarter Horse: First bred in 17th century Virginia, the oldest recognized American breed. Very versatile, prized for their ability to sprint over short distances. The compact, chunky horse has enormous muscled hind quarters, a short neat head with a small muzzle. They were originally bred from English running horses and the descendants of Spanish horses The Quarter Horse Registry has more than 3 million entries.

Roman Nose: Head having a distinct convex curve; characteristic of the Barb and various heavy horses.

Stud or Stallion: A sexually mature male horse, usually three years or older. Mustangs may reach full sexual maturity later than domesticated horses.

Stud Piles: No one absolutely knows the function of these deposits of stallion manure, sometimes huge heaps of dung added to by various stallions. The stud pile doesn't mark a territory, but may serve notice of a stallion's presence, like a bulletin board. Also, because horse manure contains partially digested food, young or vulnerable mustangs feed on the stud piles.

Teeth: Horses have twelve molars and six incisors; males have an additional tooth located behind the incisors. Permanent teeth are formed by six years of age; the high crowned teeth are the mark of a grazing animal, and continue to grow throughout the lifetime of a horse. The age of a horse can be determined by the condition and number of teeth.

Thoroughbred: Developed in England as a race horse. Carefully documented and selective breeding resulted in a breed having reliable size, courage, speed, and intelligence. The head is particularly refined, with no fleshiness, blending into a long, graceful neck. All modern Thoroughbreds descend from three founding stallions: the Byerley Turk, the Darley Arabian, and the Godolphin Arabian, bred with an English base stock.

Type: Horses of no fixed character of pedigree, having mixed ancestry.